TIMESCALES

University of Minnesota Press
Minneapolis | London

TIMESCALES
THINKING ACROSS ECOLOGICAL TEMPORALITIES

BETHANY WIGGIN, CAROLYN FORNOFF, AND PATRICIA EUNJI KIM
EDITORS

The University of Minnesota Press gratefully acknowledges the financial assistance provided for the publication of this book by the University of Pennsylvania and the University of Illinois.

Elizabeth Alexander, "The Dirt-Eaters" from *Crave Radiance: New and Selected Poems 1990–2010.* Copyright 1990 by the Rectors and Visitors of the University of Virginia. Reprinted with the permission of The Permissions Company, LLC on behalf of Graywolf Press, graywolfpress.org.

Published by the University of Minnesota Press
111 Third Avenue South, Suite 290
Minneapolis, MN 55401-2520
http://www.upress.umn.edu

ISBN 978-1-5179-0941-3 (hc)
ISBN 978-1-5179-0942-0 (pb)

Library of Congress record available at https://lccn.loc.gov/2020034651

Printed in the United States of America on acid-free paper

The University of Minnesota is an equal-opportunity educator and employer.

28 27 26 25 24 23 22 21 20   10 9 8 7 6 5 4 3 2 1

CONTENTS

Environmental Humanities across Times, Disciplines, and Research Practices

CAROLYN FORNOFF, PATRICIA EUNJI KIM, AND BETHANY WIGGIN

> Writing as writing. Writing as rioting. Writing as righting.
> On the best days, all three.
>
> —Teju Cole

Timescales is a coproduction of environmental researchers—academics, including scientists and humanists, as well as artists. Writing together, "rioting" across disciplines and across the following pages, might, "on the best of days," also anticipate "righting," in the homophones by fiction writer, essayist, and photographer Teju Cole, whose words we take as epigraph. *Timescales* provides a bound form to expansive discussions about how to represent and respond to planetary changes whose global scope and local variations exceed the purview of either the human or the natural sciences and are beyond the ken of any one writer. They exist at once in the quick time of the tweet, including Cole's, and of the lingering stretches needed to craft a collection of essays, including this one.

Animated by a creative pulse throughout, the rhythms in this volume are irregular, fast and slow, sometimes both. And necessarily so, for so entangled are our times that they sometimes surpass the disciplinary conventions that would write them—let alone right them. East Antarctica's Totten glacier, for example, was formed at the boundary between the Eocene and Oligocene epochs some 34 million years ago. In fall 2016, it was reported that this largest source of Antarctic ice had detached from the bedrock, melted from the bottom by ocean waters warming increasingly rapidly since the launch of the profoundly human experiment: the Great Acceleration.[1] This alarming occurrence made visible a massive temporal collision: the fast melt of ice formed over long millennia. The literally unsettling implications of this human–nature imbrication also stretch far into the future as sea level rise displaces human and other populations along coastlines. Ongoing events such as ice melt, akin to ocean acidification and species loss, are at once fast and slow, short and long, human and more.

These chilling simultaneities throw up a series of questions about *time* that the scholarly essays and creative interventions in this collection variously consider: What modes of understanding can help render multiple temporalities legible, and write-able? How might the grip of modernity's temporal regime, "the terrors of Progress," be loosened?[2] How quickly or slowly do environments toxify, adapt, transform, or heal? When will we exceed concentrations of atmospheric CO_2 that render life as we know it unsupportable? How might we reconcile the temporalities of biological, material, and social networks within a single environment?

While we humans have sometimes quickly responded to spectacular environmental catastrophes, we have more often failed to address local and global circumstances produced by imperialist structures of racism and speciesism developed in the *longue durée*. Faced in late 2016 with national election results in the United States that gave climate change deniers access to the levers of federal power, and with the rise of populism laced with an antiscience, white supremacist ideology in liberal democracies across the West and beyond, these questions seem more intractable than ever. They demand alternative and experimental modes of temporal engagement and visualization at the intersection of the arts, humanities, and natural and social sciences, and they suggest the need for more publicly engaged research. The environmental humanities, an emergent space for interdisciplinary knowledge production, is a response to the fact that these concerns are not adequately addressed by current institutional structures and may indeed be exacerbated by them. Experimenting across and within environmental ways of knowing, environmental humanities pick up the interdisciplinary charge that fuels environmental and sustainability studies and adds to them the recognition that environmental challenges from climate change to species loss are primarily cultural issues, questions of "what we value and what stories we tell, and only secondarily issues of science."[3]

We, the coauthors of this Introduction, have been talking together for several years while piloting a nascent, highly experimental environmental humanities program at the university level. With the contributors to *Timescales,* we have sought to articulate and embody a future-oriented practice of environmental humanism, all the while mindful of the difficult legacies and inhuman exclusions in each of those terms, in environment no less than in humanism. Writing together has sometimes been a riot. We have talked—and cajoled, and argued, and cried in frustration, and shrieked with laughter—about how to write and about how our writing might relate, if at all, to righting. In thinking about ecological timescales, we have inevitably thought about the temporal disciplines of our individual humanistic fields—ancient Mediterranean and Middle East (Kim), contemporary Latin America (Fornoff), and the early modern Atlantic (Wiggin). In attempting to multiply and connect open-ended forms of ecological knowing, we have also articulated practices for the environmental humanities. They are experimental, devised performances, inspired by our collaborations on *Timescales* with artists with whom we made several etudes in preparation for *A Period of Animate Existence,* a hybrid musical and devised performance that had its world premiere in Philadelphia in September 2017. The names of the volume's sections, "variations," "etudes," and a "coda" bear witness to our sustained "chitchats" with artists whose practices arise in dialogue with the European classical music tradition. We borrow the term "chitchat" from Jason Bell and Frank Pavia's opening chapter in the first variation. Another chitchat, between Wiggin and theater director Dan Rothenberg, concludes the volume's second interlude, recording just one of the thousands of conversations between the artists and academics who collaborated to create *Timescales.* And a final chitchat gives us, the editors, time and space to say goodbye to the period of this volume's making amid wonderfully animated collaborations.

The volume's origins lie most immediately in a conference held in 2016 that included scholarly papers, a theatrical etude, and an installation that aimed to marry art and science so as better to think through ecological timescales—including, but also before and beyond, the time of humans.[4] The conference was the brainchild of student fellows in our environmental humanities program, and the program itself was born of students' desire for sustained environmental inquiry across disciplines.[5] *Timescales,* coedited by two former program fellows and the program's faculty director, carries forward the generative conversations born of the "distant" interdisciplinarity that program students and faculty cultivate. That is, in addition to promoting dialogue between near-neighbor fields—such as literary studies and art history, or history and archaeology—*Timescales* threads connections across farther-flung fields, its pages spanning English and chemical oceanography, anthropology and geophysics. Spinning webs across what C. P. Snow long ago but no less accurately called "the two cultures" requires us to move inquiry beyond conventional disciplinary arrangements, and toward research methods and pedagogies that promote lateral thinking as much as deep disciplinary expertise.[6] Distant disciplinary collaboration can challenge unspoken assumptions grounding disciplinary cultures; it therefore requires time and trust. The going can be slow. But perhaps it is in slowness that lies the preservation of worlds.[7]

Timescales has less to do with a defined field of facts of climate change than it does with anthropogenic ecological crisis as a matter of concern.[8] To make something a matter of concern is to mobilize what Donna Haraway calls "tentacular thinking" in our approach to the pernicious effects of climate change and environmental crisis.[9] The urgency of anthropogenic climate change suggests that it is no longer sufficient for environmental scholarship to imagine that we might merely describe our complex relationship to the natural world, "to write it." Jeffrey Jerome Cohen's conceptualization of "Long Ecology" acknowledges this "more-than-human temporal and spatial entanglement" to be "an affectively fraught web of relation that unfolds within an extensive spatial and temporal range, demanding an ethics of relation and scale."[10] In other words, this state of urgency compels additional responses, some not in conventional academic writing, "to right it." The dispassionate voice of evidence-based inquiry can be heard in these pages; and so too emotional, sometimes elegiac, sometimes enraged voices.

With this volume, we aim to engage a variety of audiences by placing more traditional scholarly essays alongside explicitly experimental sections—visual modeling, storytelling, and performance documentation. They suggest alternative forms of communication and storytelling that can be adopted into and so adapt disciplinary vernaculars to the exigencies of climate change. Through the simultaneous exploration of an array of research practices, including creative and social practice arts, we can make multiple stories and formulations of environmental time audible, visible, and actionable. *Timescales* aims to translate work across different fields to promote the development of collaborative research models. By bringing together researchers employing diverse disciplinary methods, we take up Haraway's suggestion to grow a hot pile of compost.[11]

The term "Anthropocene" and its widespread adoption over the last two decades demonstrate first and foremost the need and the desire for conceptual tools to represent what modernity would call the "simultaneities of the non-simultaneous," from precipitous rates of ice melt at the poles to storms "on steroids" in the tropics.[12] Coined in 2000 by atmospheric chemist Paul Crutzen and biologist Eugene Stoermer, the term attempts to register an apparent contradiction in terms.[13] On a planet noted for its biodiversity, the term "Anthropocene" provocatively names an entire geological epoch after a single life form, the human species. The attempt has been widely criticized on a variety of grounds. As philosopher Kathleen Dean Moore notes, "We don't name new epochs after the destructive force that ended the epoch that came before." She proposes a number of other names, from the "Unforgiveable-crimescene" to the "Obscene." "If we name it after the layers of rubble that will pile up during the extinction of most of the plants and animals of the Holocene—the ruined remains of so many of the living beings we grew up with, buried in human waste—then we are entering the Obscene Epoch. It's from the Latin: ob- (heap onto) and -caenum (filth)."[14] What the "right" name might be for this time filled with wrongs has occupied others too.[15] However, as feminist philosopher and historian of technoscience Donna Haraway cautions, "Right now, the earth is full of refugees, human and not, without refuge. So, I think a big new name, actually more than one name, is warranted."[16]

Staying with the troubled term a moment longer, we find "the Anthropocene" useful in that it is marked by paradox and incommensurability. As scientists Simon Lewis and Mark Maslin note, "Although many people use the Anthropocene as a synonym for climate change or global environmental change, it is much more than these critical threats."[17] For centuries, if not millennia, before the advent of petromodernity in Stephanie LeMenager's well-phrased term, humans have been changing the planet.[18] Our species' impacts "run deeper than just our use of fossil fuels," and "so our responses to living in this new epoch will have to be more far-reaching."[19] To echo historian Dipesh Chakrabarty, we understand the Anthropocene to name an epoch rife with ironies often cruel.[20] In Chakrabarty's words, researchers of the Anthropocene must "bring together intellectual formations that are somewhat in tension with each other: the planetary and the global; deep and recorded histories; species thinking and critiques of capital."[21] Research on, and perhaps more generally research in, the Anthropocene, in other words, suggest the need for a chemist to work together with a historian, an earth scientist with a literary scholar—or at the very least to talk, read, and converse with one another. *Timescales* documents a few such interdisciplinary conversations so as to invite other "riotous" discussion and open still more avenues of inquiry.

Literary critic Rob Nixon's felicitous phrase "slow violence" marks another attempt to navigate the troubling paradoxes and simultaneities of our time. While aesthetic theory has more often considered violence produced with a flash and a bang, "slow violence" would attune our senses to the seepages and long-term leakages of environmental degradation that are often silent (or silenced) in environmental justice communities. It also foregrounds the silences in archives, including the rock layers from which scholars of the Anthropocene draw their evidence. The need for such terms that work on multiple historical registers, playing across multiple timescales, whatever their success, can be dis-spiriting. Like "slow violence," the "Anthropocene" leaves us uneasy; the times are out of joint. *Timescales* responds to these paradoxical specters that haunt us today.

Planetary time has always been bumpy, even before the Anthropocene. Déborah Danowski and Eduardo Viveiros de Castro point out that thinking about "the end" necessarily pulls us in multiple temporal directions, "Every thought of the end of the world [. . .] poses the question of the beginning of the world and that of the time before the beginning, the question of *katechon* (the time of the end, that is, the time-before-the-end) and that of the *eschaton* (the end of times, the ontological disappearance of time, the end of the end)."[22] The absence of what once was points toward finality; it demarcates a limit that cleaves the future from the past. Without the conditions that have allowed the human species to flourish, what will be our duration in time?

The ongoing ecological crisis brings the longer timescales of the planet into view, a history that began without us and will long outlast us. William Connolly warns that to attribute disruptive forces only to human agency or to treat the environment as "composed of stable patterns with only gradual change, or a set of organic balances in itself" is to disregard the long history of volatile processes of planetary transformation.[23] In the words of the collaborative of artists who made *A Period of Animate Existence,* also authors in this volume, "Something always was; something else will be; let that set you free, let that set you free."[24] Nevertheless, on a human scale, questions of finality and futurity race to the fore. The ecological crisis thus scrambles twin assumptions at the heart of Western positivism: (1) time is a linear, uninterrupted march toward progress; and (2) nature is an atemporal, boundless resource underpinning, but largely separate from, the human historical experience. Today, belief in this modern myth—of environmental continuity or harmonious evolutionary cyclicality—is shattered by events such as amphibian extinction or glacial melt.

Our choice of the word *timescales* to organize the constellation of work gathered in this book registers our desire to foreground the deep time (liveliness, experience, agency) of nonhuman processes. By insisting on the plurality of scales, and on their overlaps and entanglements, we want to push back against the discrete, measurable time period. Time cannot be measured in so many "coffee spoons."[25] Timescales rejects man as the measure of all things. Unlike a timeframe, in which a period of time is neatly bracketed from what came before and what will follow, a timescale implies depth. We envision a timescale not as a smooth slice of neatly separated layers of time but rather as composed of jostling and unstable temporalities, defined by processes of assembling and unravelling, ruptures and contingency. Thinking through timescales can also illuminate alternative viewsheds and suggest historical observations at a variety of resolutions and perspectives. Timescales understand temporality as a simultaneously material and discursive premise with its own weight or mass, modeled in ways that disclose certain kinds of information while diminishing other perspectives. In the words of art historian George Kubler:

> Time has categorical varieties: each gravitational field in the cosmos has a different time varying according to mass. [. . .] When we define duration by span, the lives of men and the lives of other creatures obey different durations, and the durations of artifacts differ from those of coral reefs or chalk cliffs, by occupying different systems of intervals and periods. The conventions of language nevertheless give us only the solar year and its multiples or divisions to describe all these kinds of duration.[26]

Here, Kubler acknowledges the different durations of time across species and objects, as well as places and spaces. With this in mind, timescales foreground a mode of critical attunement to ecological change that encompasses expansive categories like landscapes, species, class, and race, interrogating their interconnectedness and how they take shape at different rates.

In any case, what we have *not* produced is a general theory of the Anthropocene, or even a set of theses about it. For we contend, again following Rob Nixon, "We may all be in the Anthropocene but we're not all in it in the same way."[27] Just who are "we" in the Anthropocene? Who are the authors of this history? Chakrabarty's "Theses" warned that "unlike in the crises of capitalism, there are no lifeboats here for the rich and the privileged."[28] But what is the boat that is sinking? It is "Asia's historical experience," Amitav Ghosh argues, that "demonstrates that our planet will not allow these [carbon-intensive] patterns of living to be adopted by every human being." Yet today, he continues, "having entered this stage, [Asia] is trapped, like everyone else."[29] But of course "we" got here on very different ships, some of them slavers.[30]

Shipborne imperial trade floats at the origins of what sociologist Ulrich Beck called world risk society. Risk's two faces showed themselves "starting with intercontinental merchant shipping" and the risk contracts that underwrote the thousand ships that launched the global capitalist system.[31] But far from remaking social stratification dominated by class, as Beck imagined, environmental risks, including vulnerability to climate change, have reinforced existing divisions of racialized capitalism.[32] Dominated by the short-term horizon of quarterly profits, risk society has failed to manage the long-understood, long-term risk caused as humans continue to increase greenhouse gas levels in the earth's atmosphere. Superstorms, extreme weather events, and land subsidence displace human populations at an increasingly rapid clip.

We are too long overdue for, in Beck's words, a "non-nostalgic critical theory" that could "reconceptualize the past of modernity from the standpoint of the threatened future."[33] Climate scientists and journalists alike report suffering from traumatic stress disorders; even the cheerful Science Guy has been diagnosed, in *Bill Nye's Global Meltdown*, with eco-anxiety.[34] Whether this is pre- or posttraumatic stress is unclear.[35] Our changing climate is not only remaking the future planet, it's also profoundly changing how we understand our past. As climate change continues to warp time's arrow, we need more than the critical theory for which Beck called. We sense there is a groundswell of what Beck reader Wendy Hui Kyong Chun calls "new associations between knowing and doing."[36] At their best, environmental humanities aim to foster such new associations, opening knowledge production and nurturing the "right to research."[37]

Scientists are telling stories; humanists are doing experiments. Volume author Wai Chee Dimock elsewhere proposes the term "climate humanists."[38] The edge between the environmental humanities and science communication is growing. Even as climate change is scrambling the times, it is also mixing up how we apprehend and address them. Philosopher Jonathan Lear captures some of these temporal paradoxes in the concept of radical hope developed in dialogue with the history made by Crow Indian Chief Plenty Coups. In asking "for what may we hope?" Lear is also talking with Kant; rather than pose his question a priori, however, Lear wants "to consider hope as it might arise at one of the limits of human existence." Plenty Coups, Lear writes, "responded to the collapse of his civilization with radical hope." What would it mean to respond a posteriori with hope? Lear continues:

> What makes this hope *radical* is that it is directed toward a future goodness that transcends the ability to understand what it is. Radical hope anticipates a good for which those who have

the hope as yet lack the appropriate concepts with which to understand it. What would it mean for such hope to be justified?[39]

The open-ended experiments in *Timescales* are inspired by this account of Plenty Coups's radical hope, and they are offered in anticipation of a future good we can only fail to grasp. We gesture toward future goodness, writing amid increasing climate chaos. We will have written, not knowing of course if our hope will have been justified, or if our writing will have righted anything at all. This use of the future past owes an obvious debt to philosopher Rosi Braidotti. Her words propel our work across *Timescales*, "Posthuman ethics is about the pursuit of the unrealized potential of complex assemblages of subjects, at a time when the future seems rather to shrink dramatically."[40] And, as is appropriate for a coda, we will have returned to Braidotti in ours.

Timescales' eight chapters are divided across three sections, interlaced with three experimental sections. The contributions provide an array of collaborative pedagogical methods and transgressive pedagogies. Together they document a collective attempt to respond, as researchers working in a range of fields across the arts and sciences, to times at once fast and slow, and ever more alarming. To organize them, we adopt terms from the musical tradition to embrace heterogeneous scales, tempos, and modes of composition. Three sections of essayistic and scholarly chapters, or *variations,* are punctuated by three experimental artistic interludes, or *etudes. Variation* suggests a multitude of potentially transformative possibilities for scholarship and engagement, rather than one concrete answer for the Anthropocene's temporal challenges.

The first section, "Variations and Methods," begins with a chapter cowritten by literary critic Jason Bell and chemical oceanographer Frank Pavia. In "Time Bomb: Pessimistic Approaches to Climate Change Studies," Bell and Pavia seek allegorical relations between scientific and humanistic fields of inquiry in order to posit ethical responses to climate change. By advocating for pessimism, their chapter grapples with the problem of thinking deep time across the sciences and humanities, while confronting the limits of such collaborations. They promote open-ended chitchats as a practice of transdisciplinary research and writing, a mode of inquiry that does not necessarily expect a positive outcome. The second chapter, "Earth's Changing Climate: A Deep-Time Geoscience Perspective," by Jane E. Dmochowski and David A. D. Evans, both geoscientists, also takes up the call for open-ended inquiry. While some rates of ancient climate change can be known with relative precision, the event that is most comparable to today's global warming, the Paleocene–Eocene Thermal Maximum, is so ancient (some 56 million years ago) that we only have patchwork knowledge to explain it. Evans and Dmochowski assert that we need to recognize disciplinary limits and reframe scientific discussions alongside conversations in the human sciences. The third chapter, by landscape archaeologist and architectural historian Ömür Harmanşah, proposes methods from landscape archaeology as alternatives for grappling with both long and short timescales and multiple histories. Harmanşah recasts the porosity of history and indeed the Anthropocene itself through his formulation of "percolating time," drawing attention to the unevenness or leakiness of the deep and historical past. Here, archaeological methods not only highlight the failure of grand historical narratives but also act as tools for environmental justice and political activism.

The volume's first experimental interlude, Etude 1, centers on *A Period of Animate Existence,* a symphonic theater hybrid that offers meditations by elders, children, and machines on life and planetary cycles. Developed as a collaboration between Pig Iron Theatre Company's artistic director and cofounder Dan Rothenberg, composer Troy Herion, and designer Mimi Lien, all Penn Program in Environmental Humanities resident artists in 2016–17, *PAE* provides "hallucinatory visions" of the future of the planet. The materials included in *Timescales* offer glimpses into the devised performance methods employed to create a work of art with a temporal structure that is itself a variation on a classical symphony in five movements. An introduction by theater scholar and artist Marcia Ferguson contextualizes its methods within a longer history of theater making. The artists' narrative and visual contributions are bookended by a short interview about emotion and ethics between *PAE*'s dramaturge, Wiggin, and its director, Dan Rothenberg. The devised performance techniques and collaborative methods that created *PAE* echo and rebound upon the methods for environmental humanities *Timescales* puts into practice.

The volume's second set of variations turns explicitly to temporal shifts. The rhythm and pace of environmental change has become unsettled and the rate of change is increasingly accelerating. Yet in spite of clear signs of this escalation, the confounding variability of such changes renders them both too fast and too slow to mobilize effective response. Attuned to the catastrophic, we tend to overlook the gradual, incremental violence of less visible processes like ocean acidification. Conversely, faced with the escalation of record heat waves, droughts, hurricanes, and wildfires, the abnormal becomes routine, and we grow complacent. The chapters in this second section thus ask: How do we represent, manage, or react to this new era of "metatemporal instability"?[41] How might we zoom out and see the larger picture when human memory seems firmly affixed to the short term?

Charles M. Tung's contribution, "Time Machines and Timelapse Aesthetics in Anthropocentric Modernism," considers the aesthetic tools at our disposal. Looking back at the era of literary modernism, he discusses the invention of speculative techniques including the time machine and time lapse as strategies for depicting non-events, hyperobjects, and processes that unfold across more-than-human timescales. Such strategies, Tung proposes, problematize human periodicity by representing "scalar misalignments" and estranging "earthly temporal units." Yet he cautions that they also problematically smooth over disjunctures, rendering the passage of time fluid and seamless. In "Fishing for the Anthropocene: Time in Ocean Governance," Jennifer E. Telesca addresses how time is parceled from the perspective of ethnographic anthropology and takes up the international bureaucratic regulation of marine life. In the name of "conservation," she argues, time has become a site of instrumentalization, a tool to be manipulated in the market-driven race against finitude and extinction. Telesca documents how "technocratic time"—which is forward thinking, linear, unidirectional, and irreverent of the past—enables the commodification of fish as stock, stripped of agency and set outside history.

In Etude 2, the second experimental interlude, artist Mary Mattingly writes about *WetLand*, a utopian experiment in sustainable living, suffused with radical hope. Mattingly was PPEH's artist in residence in 2015–16 and her *WetLand* boat was at the heart of *The WetLand Project*, which concluded in June 2017. The boat and art installation as well as the social practices that kept it afloat intended to spark conversations and new collaborations to address the problems we face in an era of rising sea level. *WetLand* was docked on the banks of Philadelphia's Schuylkill River at Bartram's Garden, America's oldest botanical garden. Today, the Garden sits nestled among residential neighborhoods and what was the East Coast's largest oil refinery until it exploded in June 2019. The friction of garden pastoral with industrial sublime inspired the *Project,* and throughout the spring and early summer 2017, it seeded other tools for *WetLand,* subject of landscape architect Kate Farquhar's contribution to this etude. While awaiting tow to its next home in late summer 2017, the boat took on water and sank; heavier, "unseasonal" rains point to the city's warmer and wetter "new normal."[42] The boat had to be floated and removed, prompting Farquhar's meditations on collaborative endings and beginnings.

The penultimate section, "Repetitions and Variations," delves into the temporal collapse of past and present in the Anthropocene, while addressing the impact that environmental destruction has had on human and nonhuman communities. In spite of its seeming "newness," the irreversibility of environmental ruin compels us to look back in order to move forward. As noted by Haraway, this sort of looking back, or mourning, "is intrinsic to cultivating response-ability."[43] The chapters in this section discuss the stakes of environmental and cultural loss as well as attempts at remediation across different landscapes, both past and present.

Wai Chee Dimock's "Vanishing Sounds: Thoreau and the Sixth Extinction" explores extermination through sound, or its absence. Via a reading of Henry David Thoreau, she considers what is at stake in mobilizing political categories for animals and in deploying naturalized categories for some humans. By exploring the loss of biodiversity as a sonic phenomenon, Dimock foregrounds the sensorial, affective register of extinction. In "Hoopwalking: Human Rewilding and Anthropocene Chronotopes," Paul Mitchell explores contemporary practices of human rewilding in the North American West as a program of restoration ecology that builds future imaginaries by engaging with (re)imagined pasts. Mitchell situates such rewilding movements within the context of settler colonialism while espying in the rewilders "on the hoop" practices of multispecies entanglement and cohabitation. Iemanjá Brown's final chapter, "Dirt Eating in the Anthropocene," describes geophagy, or the desire to consume dirt, in poetic and personal terms. Here, geophagy would bring the earth's deep time and possible futurities into the present through the body. Reflections on her own hunger for dirt leads to a discussion of contemporary poet Elizabeth Alexander's "Dirt Eaters," which traces geophagy as a subversive tool used by enslaved black women to foster intimacy with the ground and their social and material histories. Historical and poetic encounters with geophagy bring issues of labor, production, and subjectivity to bear on Anthropocene scholarship.

In *Timescales'* concluding etude, the Los Angeles–based Salvadoran artist Beatriz Cortez presents speculative work about futurity. Cortez imagines alternate futures, perhaps even futures in which humans are no longer able to inhabit Earth. Yet even such radically divergent fates do not erase human difference or cultural memory. Cortez's "Memory Insertion Capsule" is a spaceship fashioned with steel lumps that look like river rocks, evoking Indigenous construction techniques. Cortez observes, "We always imagine indigenous people being part of our past. I wanted to imagine [them] as part of our future."[44] Cortez's etude reflects on simultaneity: the way in which the past informs the present and the future. To think with climate change is to think across timescales, to simultaneously engage with mass migration and alienation, and their connections to brutal and racist colonial pasts.

Cortez's concluding etude points to the foresight of Indigenous cosmologies when it comes to the current crisis. It echoes the observation of a Yanomami shaman interviewed by Danowski and Viveiros de Castro, who remarked, "Whites are not afraid of being crushed by the falling sky as we are. But one day they will be, maybe as much as we are!" To which Danowski and Viveiros de Castro write, "This day is apparently dawning."[45] Whether or not the Subcommission on Quaternary Stratigraphy's recommendations regarding the scientific usefulness and definition of the Anthropocene are fully accepted by the International Union of Geological Sciences, the pervasive sense that something is missing foregrounds the urgency and the commitments of this present volume.[46]

Timescales is not unlike a manifesto—if a manifesto could be made proper to an age that has lost its faith in reason's progress. The manifesto, born of the nineteenth-century's militant optimism about the march of history, feels rather like an ancient relic. And yet, it is also not, for all our sympathy with Hamlet, a tragedy. Perhaps it is a manifesto "meet / to put an antic disposition on."[47] In any case, it bears the marks of radical hope.

In addition to new alliances in discipline and thought, the contributions to our volume experiment with the different temporalities of scholarship and writing, toggling between the unwieldy urgency and the long-term implications of both climate change and academic work. For some, the tweet, the mode by which our epigraph was produced, responds audibly and quickly to multiple audiences. The immediacy of the tweet in some ways echoes the act of teaching in that it performs idea sharing and collaborative thinking in real time. Like the riot, which often produces collective chants, phrases, and hashtags, the tweet effectively expresses frustration and the need to respond in a visceral or easily digestible medium. Rioting pushes an important release valve; flaring up and dying down, it constitutes a vital aspect of the slower process of righting.

Alongside more immediate responses, our volume recognizes and advocates slower approaches to scholarship. While it might seem counterintuitive to respond to the acceleration of climate change with a politics of deceleration, in the words of Isabelle Stengers, "Slow does not mean idle."[48] While ecological crisis is urgent, Stengers warns that we must tread cautiously when thinking about the "urgency" of response, lest it be mobilized to reproduce business as usual: "universal" solutions cloaked in a rhetoric of corporate sustainability that perpetuate the very same inequities of capitalism. To counter this threat Stengers espouses "slow science." Analogous to the slow food movement, slow science privileges quality over quantity. To make such a pivot is to embrace protracted timescales: to accept downtempo or longer production times over the instantaneity of ready-made solutions. The slowing down advocated by Stengers involves acts such as paying attention, prolonging and hesitating— precisely the types of bearings on seas for which capitalism offers no charts.[49]

This new kind of scholarship demands a shift from heroic ambitions and grand theory to a more modest scale that focuses on an ethic of care. We want to amplify Jenny Price's admonition to "stop saving the planet," and instead, following the advice of the *Dear Climate* project, "stay close to home."[50]

Rioting and righting climate change start at home, in each of our lives and in the maintenance of steady relationships sustained over time.[51] Such perspectives scale up the importance of habit and the unremarked practices of maintenance, often homely in origin. Here we take inspiration from Mierle Laderman Ukeles's "Manifesto for Maintenance Art." In her manifesto, Ukeles incisively asks, "After the revolution, who's going to pick up the garbage on Monday morning?"[52] Ukeles's performances of the late 1960s urged the recognition of tedious and slow forms of labor, too often carried out by women in the shadows. *Timescales*' three editors are cis-women scholars/ colleagues/mentors/students, who have figuratively and literally picked up the trash and taken out the recycling while managing environmental humanities programming and their professional relationships with each other for the past five years. Institutionalizing these sorts of experimental spaces for programming and collaboration requires administrative ingenuity and maintenance, labor that within academia too often goes unnoticed and unrewarded. This labor is akin to housework; maintenance that has the potential to be transformative. As Sara Ahmed writes, "Feminist housework does not simply clean and maintain a house. Feminist housework aims to transform the house, to rebuild the master's residence."[53] Nonetheless (even among feminists) collaborative work can be challenging; disruptive frictions inevitably emerge. Writing, rioting, and righting are practices that require perseverance in the face of such friction, which is to say, reliability, open-mindedness, fairness, and flexibility.

In the spirit of unexpected collaborations and open-ended outcomes, *Timescales* is a composition of the shared conversations and productive misunderstandings that emerged from the active, intimate, and careful efforts of relating. It attempts to model an open, public knowledge commons that prioritizes engagement across generations and disciplines in order to nurture learning communities that did not previously exist. Likewise, each of our contributors proposes alternatives for scholarship that foster horizontality within the traditionally vertical chains of knowledge production. Open-ended outcomes are valuable because they generate new questions without necessarily expecting solutions. Solving the problems of

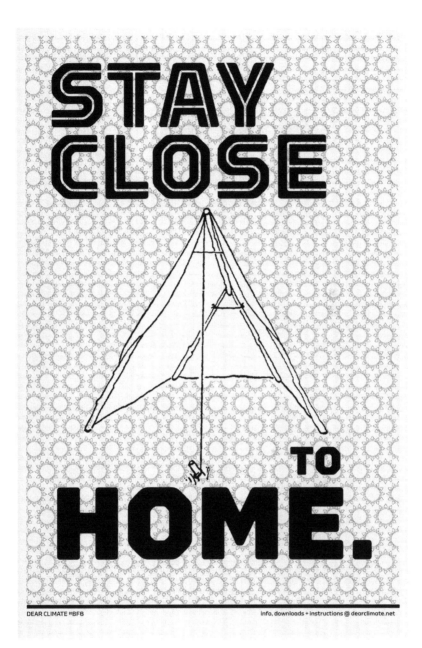

FIGURE I.1. "Stay Close to Home" is one of many posters from the *Dear Climate* project, exhibited in 2014 and created by Una Chaudhuri, Fritz Ertl, Oliver Kellhammer, and Marina Zurkow. Courtesy of www.dearclimate.net.

the Anthropocene is unlikely—but to understand them and begin to transform them, we need to embrace experimentation and productive failures. The creation of spaces for risky conversations and collaborative experimentation reworks institutional mechanisms of exclusion and begins to break down disciplinary boundaries and foster a spirit of reciprocity among thinkers. As Zoe Todd has urged, "Reciprocity of thinking requires us to pay attention to who else is speaking alongside us. It also positions us, first and foremost, as citizens embedded in dynamic legal orders and systems of relations that require us to work constantly and thoughtfully across the myriad systems of thinking, acting, and governance within which we find ourselves enmeshed."[54] To think with the Anthropocene, Bruno Latour explains, is not to turn "to nature" but to "probe on the near side" of it, to translate what needs to be done and "begin to treat our madness"[55] or as Amitav Ghosh puts it, our "great derangement."[56] Through *Timescales'* attempts to think with others, we hope to create spaces within the academy for conversations and experiments that may productively lead nowhere.

NOTES

1. "Welcome to the Anthropocene," http://www.anthropocene.info/great-acceleration.php, accessed September 10, 2017.

2. Heather Swanson, Anna Tsing, Nils Bubandt, Elaine Gan, "Introduction: Bodies Tumbled into Bodies," in *Arts of Living on a Damaged Planet* (Minneapolis: University of Minnesota Press, 2017), M7.

3. Ursula Heise, *Imagining Extinction: The Cultural Meanings of Endangered Species* (Chicago: University of Chicago Press, 2016), 5.

4. "Timescales," https://timescalesconference.wordpress.com, accessed September 21, 2017. To date, the installation, *Date/um,* has been exhibited three times. The installation is documented at https://ppeh.sas.upenn.edu/experiments/dateum, accessed August 26, 2020.

5. "Penn Program in Environmental Humanities Manifesto," http://www.ppehlab.org/manifesto/, accessed September 21, 2017.

6. C. P. Snow, *The Two Cultures and the Scientific Revolution* (New York: Cambridge University Press, 1959).

7. The challenge, as we and so many others are asking, is this: how to represent and to respond to the "slow violence" that is the seeping signature of the Anthropocene, polluting discrete strata and corporeal bounds. Rob Nixon, *Slow Violence and the Environmentalism of the Poor* (Cambridge, Mass.: Harvard

University Press, 2013). We return to this challenge in the Coda to this volume, turning more squarely to Isabelle Stengers's "Manifesto for Slow Science," in *Another Science Is Possible*, trans. Stephen Muecke, 106–32 (Cambridge, UK: Polity, 2018).

8. Bruno Latour, "Why Has Critique Run Out of Steam? From Matters of Fact to Matters of Concern," *Critical Inquiry* 30, no. 2 (Winter 2004): 225–48.

9. Donna Haraway, *Staying with the Trouble: Making Kin in the Chthulucene* (Durham, N.C.: Duke University Press, 2016), 31.

10. Jeffrey Jerome Cohen, *Stone: An Ecology of the Inhuman* (Minneapolis: University of Minnesota Press, 2015), 41.

11. Haraway, *Staying with the Trouble*, 57.

12. Marxist philosopher Ernst Bloch elaborated the "simultaneities of the non-simultaneous" in *Inheritance of the Time (Erbschaft dieser Zeit)*: "Older times than the modern ones continue to have an effect in older strata . . . [and] contradict the Now; very strangely, crookedly, from behind." Translated by Frederick J. Schwartz in his article "Ernst Bloch and Wilhelm Pinder: Out of Sync," *Grey Room* 3 (Spring 2001): 58.

"For some of the physical processes discussed here, one can view increasing carbon dioxide in the atmosphere as steroids for the storms." "What We Know about the Climate Change–Hurricane Connection," accessed September 10, 2017, https://blogs.scientificamerican.com/observations/what-we-know-about-the -climate-change-hurricane-connection/.

13. The two scientists proposed the term in 2000, in the pages of the *Newsletter of the International Geosphere-Biosphere Programme*. The article is reprinted in the October 31, 2010, issue of the IGBP's *Newsletter*, with some historical contextualization and references to other important formulations of the Anthropocene published by Crutzen in the early years of the new millennium. "Have we entered the 'Anthropocene'?," http://www.igbp.net/news/opinion/opinion /haveweenteredtheanthropocene.5.d8b4c3c12bf3be638a8000578.html, accessed September 10, 2017.

14. Kathleen Dean Moore, "Anthropocene Is the Wrong Word," *Earth Island Journal*, Spring 2013, http://www.earthisland.org/journal/index.php/eij/article /anthropocene_is_the_wrong_word/, accessed September 10, 2017.

15. See, for example, Jason W. Moore, ed., *Anthropocene or Capitalocene? Nature, History, and the Crisis of Capitalism* (Oakland, Calif.: PM Press, 2016).

16. Donna Haraway, "Anthropocene, Capitalocene, Plantationocene, Chthulucene: Making Kin," *Environmental Humanities* 6 (2015): 160.

17. Simon L. Lewis and Mark A. Maslin, *The Human Planet: How We Created the Anthropocene* (New Haven, Conn.: Yale University Press, 2018), 6.

18. LeMenager defines petromodernity as "modern life based in the cheap energy systems made possible by oil." Stephanie LeMenager, *Living Oil: Petroleum Culture in the American Century* (New York: Oxford University Press, 2014), 67.

19. Lewis and Maslin, *The Human Planet*, 6.

20. On irony as the figure of the Anthropocene, see also Bethany Wiggin, "The Germantown Calico Quilt," in *Future Remains: A Cabinet of Curiosities for the Anthropocene,* ed. Gregg Mitman, Robert Emmett, and Marco Armieri, 149–58 (Chicago: Chicago University Press, 2018).

21. Dipesh Chakrabarty, "The Climate of History: Four Theses," *Critical Inquiry* 35, no. 2 (Winter 2009): 213.

22. Déborah Danowski and Eduardo Viveiros de Castro, *The Ends of the World,* trans. Rodrigo Nunes (Cambridge, UK: Polity, 2016), 19.

23. William Connolly, *Facing the Planetary: Entangled Humanism and the Politics of Swarming* (Durham, N.C.: Duke University Press, 2017), 92.

24. Will Eno, Troy Herion, Mimi Lien, and Dan Rothenberg, *A Period of Animate Existence,* Movement 2, in performance.

25. "For I have known them all already, known them all / Have known the evenings, mornings, afternoons, / I have measured out my life with coffee spoons." T. S. Eliot, "The Love Song of J. Alfred Prufrock," *Poetry Foundation,* https://www.poetryfoundation.org/poetrymagazine/poems/44212/the-love-song-of-j-alfred-prufrock, accessed June 26, 2020.

26. George Kubler, *The Shape of Time: Remarks on the History of Things* (New Haven, Conn.: Yale University Press, 1962), 84.

27. Rob Nixon, "The Anthropocene: The Promises and Pitfalls of an Epochal Idea," *Edge Effects,* November 6, 2014, http://edgeeffects.net/anthropocene-promise-and-pitfalls/, accessed September 10, 2017.

28. Chakrabarty, "The Climate of History," 221.

29. Amitav Ghosh, *The Great Derangement: Climate Change and the Unthinkable* (Chicago: University of Chicago Press, 2016), 92.

30. On which ship lends an apt figure for the Anthropocene, see also Marco Armiero, "Of the Titanic, the Bounty, and Other Shipwrecks," *intervalla* 3 (2015): 50–54. The rate of survival of the *Titanic*'s first-class passengers was significantly higher than those in steerage.

31. Ulrich Beck, *World at Risk,* trans. Ciaran Cronin (Cambridge, UK: Polity, 2009), 4 and 7.

32. Heise, *Imagining Extinction,* 223.

33. Beck, quoted in Heise, 49.

34. Meteorologist and climate journalist Eric Holthaus has tweeted on many occasions about anxiety and climate change. The outpouring of sympathy after Holthaus's January 2018 tweetstorm about his own mental disorders is covered in Daniel Oberhaus, "Climate Change Is Giving Us 'Pre-Traumatic Stress,'" *Motherboard,* February 4, 2017, https://motherboard.vice.com/en_us/article/vvzzam/climate-change-is-giving-us-pre-traumatic-stress, accessed February 13, 2018. Bill Nye, *Global Meltdown,* http://channel.nationalgeographic.com/explorer/episodes/explorer-bill-nyes-global-meltdown/, accessed February 13, 2018.

35. E. Ann Kaplan, *Climate Trauma: Foreseeing the Future in Dystopian Film and Fiction* (New Brunswick, N.J.: Rutgers University Press, 2016).

36. Wendy Hui Kyong Chun, "On Hypo-Real Models or Global Climate Change: A Challenge for the Humanities," *Critical Inquiry* 41 (Spring 2015): 679.

37. Arjun Appadurai, "The Right to Research," *Globalisation, Societies, and Education* 4, no. 2 (2006): 167–77.

38. Wai Chee Dimock, "Experimental Humanities," *PMLA* 132, no. 2 (March 2017): 241–49.

39. Jonathan Lear, *Radical Hope: Ethics in the Face of Cultural Devastation* (Cambridge, Mass.: Harvard University Press, 2006), 103.

40. Rosi Braidotti, "Posthuman, All Too Human: The Memoirs and Aspirations of a Posthumanist," The 2017 Tanner Lectures (delivered at Yale University, March 1–2, 2017), 27, https://tannerlectures.utah.edu/Manuscript%20for%20Tanners%20Foundation%20Final%20Oct%201.pdf, accessed February 13, 2018.

41. Danowski and Viveiros de Castro, *The Ends of the World*, 8.

42. As journalist Brian Kahn tweeted, "Heavy rains sunk a climate change art installation. Irony is also dead." https://twitter.com/search?q=irony%20art%20climate%20change&src=typd. See also "Floating Art Installation about Rising Waters Sinks during Storm," https://weather.com/news/news/wetland-floating-art-installation-storm-pennsylvania-sunken-schuylkill-river, accessed September 21, 2017.

43. Haraway, *Staying with the Trouble*, 38.

44. Jori Finkel, "For Latino Artists in Sci-Fi Show, Everyone's an Alien," *The New York Times*, August 25, 2017.

45. Danowski and Viveiros de Castro, *The Ends of the World*, 74.

46. "Working Group on the 'Anthropocene,'" https://quaternary.stratigraphy.org/workinggroups/anthropocene/, accessed September 19, 2017. See also Maya Lin's digital monument to species extinction, *What Is Missing*, https://whatismissing.net, accessed September 24, 2017.

47. William Shakespeare. *Hamlet*, I.5.172–73.

48. Isabelle Stengers, *Another Science Is Possible! A Manifesto for Slow Science* (Cambridge, UK: Polity, 2018).

49. Isabelle Stengers, *In Catastrophic Times: Resisting the Coming Barbarism*, trans. Andrew Goffey (n.p.: Open Humanities Press and meson press, 2015), 51 and 105.

50. Jenny Price, "Stop Saving the Planet!" March 29, 2012, https://www.sallan.org/pdf-docs/StopSavingPlanet.pdf.

51. We have been inspired by the open-ended *Dear Climate* project exhibited in 2014 and created by Una Chaudhuri, Fritz Ertl, Oliver Kellhammer, and Marina Zurkow. Publicly and freely available online, it offers tools to foster "inner climate change." http://www.dearclimate.net/#/homepage, accessed September 15, 2017.

52. Mierle Ladermann Ukeles, "Manifesto for Maintenance Art 1969! Proposal for an Exhibition 'CARE.'" Originally published in Jack Burnham, "Problems of

Criticism." *Artforum*, January 1971, 41; reprinted in Lucy Lippard, *Six Years: The Dematerialization of the Art Object* (New York: New York University Press, 1979), 220–21.

53. Sara Ahmed, *Living a Feminist Life* (Durham, N.C.: Duke University Press, 2017), 7.

54. Zoe Todd, "An Indigenous Feminist's Take on the Ontological Turn," *Journal of Historical Sociology* 29, no. 1 (2016): 19.

55. Bruno Latour, *Facing Gaia: Eight Lectures on the New Climatic Regime* (Cambridge, UK: Polity, 2017), 20.

56. Ghosh, *The Great Derangement*.

PART I

VARIATIONS AND METHODS

CHAPTER 1

Time Bomb

Pessimistic Approaches to Climate Change Studies

JASON BELL AND FRANK PAVIA

Nuclear Testing and Oceanography

FRANK PAVIA: I'm an oceanographer but I've never seen below the top five meters of the ocean. I've never seen the ocean. I've been on boats that have traveled across the ocean, collecting samples from the bottom of the ocean, and I've never experienced or seen below the surface.

JASON BELL: You've never seen the thing that you use as your index or marker, either.

FP: No, I've never seen an isotope of protactinium. I've seen a solution that has protactinium in it, but the only way that I know it has protactinium in it is by measuring the protactinium atoms.

JB: As a literary critic, I don't even know if I'm asking the right question. We've spent so much time listening to the Clash and watching movies from the 1970s and thinking about how humans have represented their attempt to represent the ocean, so I feel like asking, "what does surf punk tell us about bomb radiocarbon and radioisotopes?" But we don't really want to be asking that question.

FP: That's a pointless question because it leads us to a predetermined outcome.

JB: Let me start then by asking whether you learned anything about oceanography from the process of writing this paper, not just from listening to surf punk or watching *Apocalypse Now*.

FP: I'm not sure how much I learned about oceanography, but I learned a lot about literary criticism and surf punk music and Francis Ford Coppola. I don't find that concerning in any way. I take it that you learned a fair amount about oceanography?

JB: I feel like I know more than the average person does.

FP: I hope you do.

JB: Aside from intellectual enrichment, what's the point of this kind of work? Or is there no point? There's no point.

FP: I find it encouraging.[1]

Radiocarbon (^{14}C), a rare, radioactive isotope of carbon with a half-life of approximately 5700 years, is formed naturally in the atmosphere by cosmic ray interactions with nitrogen. Under normal circumstances, neutrons generated by cosmic rays bombard nitrogen atoms in the atmosphere, converting them to radiocarbon by forcing the most common isotope of nitrogen (^{14}N) to emit a proton.[2] The natural budget of ^{14}C in the atmosphere is controlled by changes in solar activity, Earth's geomagnetic field, and exchange with carbon pools of the terrestrial biosphere and ocean.[3]

Human activity has disturbed the natural balance of radiocarbon, most notably through aboveground nuclear testing. Nuclear weapons are made possible by chain reactions, wherein the production of one neutron perpetuates a loop of neutron production. As a result, atmospheric nuclear testing releases enormous external fluxes of neutrons into the stratosphere. Nearly all of the neutrons generated by nuclear weapons tests react with atmospheric ^{14}N to form ^{14}C far in excess of what is produced naturally by cosmic rays.[4]

The Trinity test in 1945 marked the first detonation of a nuclear weapon. Less than a month later, the United States dropped atomic bombs over the Japanese cities of Hiroshima and Nagasaki. For the next eighteen years, more than five hundred aboveground nuclear tests were conducted before the United States, USSR, and United Kingdom signed the Limited Test Ban Treaty (1963), banning aboveground, underwater, and outer space nuclear testing. The motivations for the Limited Test Ban Treaty (LTBT) were manifold, but were focused by concerns about the dispersal and human health hazards of radioactive fallout.[5]

The effect of nuclear testing is evident in records of ^{14}C measured in ground-level air over time. Between the onset of measurements in 1955 and the LTBT in 1963, atmospheric ^{14}C/^{12}C ratios increased tenfold over pre-nuclear values. In particular, a massive spike in atmospheric ^{14}C/^{12}C was observed in 1963, corresponding to the final

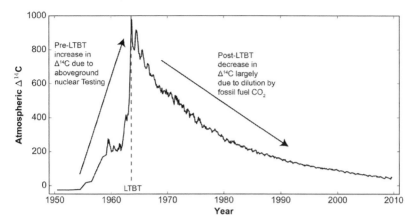

FIGURE 1.1. Northern Hemisphere atmospheric radiocarbon contents 1950–2010 in delta notation (where $\Delta^{14}C$ refers to the $^{14}C/^{12}C$ ratio of CO_2 relative to a standard, corrected for age and isotopic fractionation).[6] The dashed line in 1963 demarcates the signing of the Limited Test Ban Treaty (LTBT). The pre-LTBT increase in atmospheric $\Delta^{14}C$ is due to production from aboveground nuclear testing. The post-LTBT decline is primarily due to dilution by radiocarbon-free fossil fuel CO_2 emissions, but also partly by atmospheric exchange with the oceans and terrestrial biosphere.

burst of aboveground tests before the LTBT was signed (Figure 1.1). In 1983, Reidar Nydal, one of the first scientists who sought to measure bomb-produced ^{14}C, wrote regarding his motivations, "like most people 20–30 years ago we were worrying about the possible danger to human health from the enormous nuclear activity in the atmosphere." But Nydal and his colleagues also "realized that the radioactive isotopes already injected to the atmosphere could be useful tracers in geophysical research."[7]

The 1963 ^{14}C peak tagged and time-stamped the atmosphere. By measuring ^{14}C in plants and ocean waters, scientists could track how long it took the bomb spike of ^{14}C from 1963 to enter different reservoirs within the Earth system.[8] The timescales of Earth's carbon cycle could now be established. It became possible to calculate how fast carbon could pass from the atmosphere into land plants and the ocean over periods of years to decades.

Two findings using bomb ^{14}C stand out as particularly important for understanding how the Earth will naturally respond to and partially mitigate anthropogenic climate change. By measuring inventories of bomb ^{14}C in the ocean, Stuiver was able to calculate the rate of gas exchange between atmospheric CO_2 and the ocean.[9] Twelve years later, Siegenthaler and Joos tuned a numerical model of ocean physics to reproduce the spatial distributions of both natural and bomb-produced ^{14}C.[10] Their study solved for mixing rates at different depths in the ocean and the transport rate of carbon from the ocean surface to the interior. The first study determined how fast the ocean can absorb CO_2 from the atmosphere; the second determined how fast the ocean could sequester that CO_2 at depth. Together, these two studies make up the foundation upon which scientists are able to predict the removal rate of fossil fuel CO_2 on both human (years to decades) and geologic (thousands of years) timescales, the basis for modeling projections of future climate change due to greenhouse forcing.

Cultural critics and Earth scientists alike frequently cite climate change as a unique emergency in the preservation of life and the classification of knowledge.[11] Anthropogenic climate change threatens to dissolve a barrier separating histories of the human species and culture from the natural world. Both conservative and alarmist models project such dramatic modifications to the Earth system that the reorganization of human civilization and nonhuman ecosystems over the next millennium seem inevitable.[12] Mass migration and catastrophic biodiversity loss are forecast over this extended period of time.[13] The likelihood is remote that humans will avert permanent changes to the Earth system or know in advance what those changes will be. Sea level rise, the spread of tropical diseases, and animal and plant extinctions may be unavoidable except through technological fixes (launching sulfur dioxide into the atmosphere, carbon sequestration), side effects unknown, or a revolution in the global energy economy.[14] Even if geoengineering or a total restructuring of petrocapitalism could have been implemented before 2020, the climate change already incurred may be irreversible.[15] Yet the principal response to climate change in the academy is a call for new alliances and collaborations between scientists and humanists to produce social transformation and technological solutions.[16]

To make matters worse, the difficulty of predicting medium- and long-term futures, along with the unimaginably enduring presence of greenhouse gases in the atmosphere, pose a serious dilemma for communicating the urgency of climate change to the public and conducting research on its effects. In order to explain why climate change matters and how we might begin to develop new interdisciplinary paradigms to conceptualize its likely impacts, scholars have advanced a new geologic epoch. As opposed to epochs like the Pleistocene or Holocene, the Anthropocene is marked by human activity, not geologic phenomena like the waxing and waning of ice sheets. Instead of atmospheric warming associated with industrialization, recent recommendations peg the onset of the Anthropocene to the release of nuclear fallout in the 1950s.[17] Therefore, as an index and symptom of the Earth's response to human technology, bomb radiocarbon anchors oceanographic and historical work on anthropogenic climate change and crystallizes its central problem: how to coordinate work at extreme timescales distended far beyond human lifespans in disciplines that hold different methodological norms? To rephrase that problem in this essay's terms, how can an oceanographer (Frankie) and a literary critic (Jason) collaborate without begging the question that such research is possible? If our premises hold—that the harms of climate change are unavoidable, and that the timescales of climate change and radiocarbon indexes exceed generational experience—we need interdisciplinary formations with lower confidence in and less emphasis on research outcomes.

In this essay, we invoke a line of pessimistic philosophy to explore a mode of interdisciplinary work that does not insist on the predetermined value of research outcomes or the active generation of knowledge. Extending from Voltaire through Schopenhauer, Nietzsche, and the early twentieth-century existentialists, this pessimistic tendency gravitates toward a reality constructed through unsatisfied desire and suffering, roughly speaking. Although epistemic pessimism is underdeveloped in comparison to this metaphysical and ethical articulation (the world is bad and getting worse), one can hold a pessimistic attitude toward the production of knowledge, its verifiability, communicability, or stability. Our objective, described in greater detail below, was to study the relationship between nuclear testing and anthropogenic warming from scientific and humanistic perspectives simultaneously. This task underscores the problem of epistemic pessimism that lurks behind scholarly collaboration. The coauthors each brought an object—bomb radiocarbon and surf punk music—to the table. Without making historicist, formalist, or scientific claims about the relationship of the former to the latter, we engaged in what we call "interdisciplinary pessimism." Collaborative talk, reproduced throughout the essay in dialogic asides, is both the method and outcome of our disciplinary work. These dialogues represent a form of companionability catalyzed or initiated by grief or panic. Our conversations organize what we call a passive counterculture, or oppositional attitude toward the dominant relationships linking scientists and humanists in climate studies. By conversing without hope of an instrumental outcome, we identified analogous kinds of inactivity in an unlikely place: surf punk music. The insights that follow are not valueless but rather value indeterminate. Whatever use they might have for scholars, administrators, or policymakers is entirely open-ended.

Our experimental technique does not promise success, and may in fact look like a failure to those invested in optimistic (outcome-dependent) interdisciplinarity.[18] What would success even look like under such a framework? In keeping with our method, we do not offer a conclusive answer. For us, success meant the choice to think together instead of separately, as we do on a typical day in the laboratory or library. That simple decision allowed us to explore mutually unfamiliar territory across disciplines and to take anything we found as a legitimate result. We hope to make the case for a greater diversity in interdisciplinary styles, especially when considering massive timescale phenomena like radioactive decay and climate change.

Negotiating and distributing the labor of climate change studies between disciplines like oceanography and literary criticism requires, first, a theory of what divides scientific from humanistic disciplines and, second, a theory of whether and how that division might be surmounted. Our first principle is that the sciences and the humanities are concerned with a shared collection of stuff (to use a technical term) like oceans, atmospheres, neurotransmitters, computational networks, books, characters in books, cultures, conflicts, and love. The common sphere of reality to which scientists and humanists direct their attention is prior to but accessed by observation and not apportioned into separate, inquiry-dependent domains. This is not an uncontroversial position because it is susceptible to reductive accounts of realism (described in the next section) and because it forecloses models of research based on alternative ontological preconditions; for example, pure ideality, which might structure the division of disciplinary labor differently. These different ontologies are at the

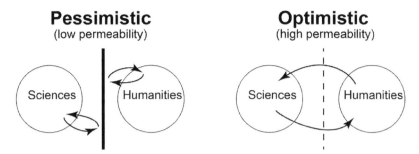

FIGURE 1.2. Schematic of interdisciplinarity as permeability.

margins of contemporary debates over climate change studies and beyond the scope of this paper with the exception of ontological pluralism, a position advanced in a series of recent papers by Jonathan Kramnick and Anahid Nersessian to discredit interdisciplinarities that subordinate humanistic to scientific explanation.[19] Ontological pluralism posits that each discipline studies a bounded part of the world to which its particular mode of explanation is suited. Close reading belongs to poems, isotopic analysis to ocean currents. But we contend that if reality appears to be carved up into niches appropriate to each discipline—Ishmael in *Moby-Dick* for literary scholars, the Pacific Ocean and its currents for oceanographers—this partition of the world follows from the particular techniques, strategies, norms, and methodologies of knowledge production constitutive of each discipline. Specifically, we distinguish between types of *falsifiable* scientific explanation, capable of being proven false through experiment, and types of *nonfalsifiable* humanistic interpretation that, while susceptible to judgments of taste (one interpretation being more interesting or penetrating or astute than another), can only be proven comparatively plausible.[20] While literary criticism does not make novels and oceanography does not make oceans, the semblance of bounded subject matter for each discipline is an epiphenomenon of this boundary in epistemology. In contrast to Kramnick and Nersessian's claim that the kernel of both scientific and humanistic inquiry is explanation, we draw a sharp distinction between scientific and humanistic epistemologies, broadly construed.

At the risk of simplification—a risk we will have to accept on account of brevity—this essay assumes that not so much has changed since C. P. Snow described the sciences and humanities in a 1959 lecture as two cultures with "literary intellectuals at one pole—at the other scientists, and as the most representative, the physical scientists. Between the two a gulf of mutual incomprehension—sometimes (particularly among the young) hostility and dislike, but most of all lack of understanding."[21] Snow might have found it strange, then, to encounter a paper coauthored by a literary critic and an oceanographer, both young, neither hostile nor completely incomprehensible to one another. Almost sixty years after *The Two Cultures*, research projects bridging the sciences and humanities seem various and robust, especially in fields like evolutionary and cognitive studies where the evident object of scientific study is the human mind or climate change studies where the object of study threatens to transform human existence. Nevertheless, Snow would be gratified to discover his original problem no less intractable: even when looking at the same thing from acute angles, scientists and humanists seem to produce different kinds of knowledge differently. Interdisciplinarity, then, is not an issue of adapting different methods to different subjects or topics, but rather the process of translating different kinds of knowledge about the same stuff from one discipline to another. The relative ease of translation or transmission across the membranes holding disciplines apart could thus be defined, to borrow from physical science, in terms of permeability. After describing available types of interdisciplinary research conducted under high-permeability conditions, we turn to a low-permeability state. We briefly explore one type of interdisciplinary pessimism, making small talk, in relation to bomb radiocarbon and surf punk and address a major criticism of this approach, its apparent absurdity.

Interdisciplinary Optimism

FP: In science, there's a general feeling that people should publish negative results more often. If you do an experiment, and you find your experiment didn't work, most of the time you don't write it up because you have a failed experiment. But it's likely that someone else is going to try the same thing, so if you don't publish, you're indirectly causing someone else to waste their time. You're actually impeding progress by not exposing your failures to the world.

JB: Interdisciplinary optimism is so focused on being successful that there's no room to fail. There's a real risk that what we're saying about bomb radiocarbon and surf punk, or trying to say, is a failure.

FP: We have no idea whether this paper will work. I hope it doesn't.

JB: Which is in the spirit of the paper.

Most collaborative research between the sciences and humanities is predicated on the smooth transfer of method and knowledge across a line drawn in the disciplinary sand, as well as a belief that collaboration produces insights comprehensible and useable in both disciplines. Optimism's stranglehold is unsurprising, since the alternative (the impossibility or unlikelihood of transmission, the futility of the result) seems to provide little reason at all to be interdisciplinary. Further, the prevailing justifications for interdisciplinary work, including the rescue of the humanities from institutional belt-tightening, improved ethics of scientific experimentation and communication to the public, and the intrinsically interdisciplinary nature of emergencies threatening human existence (like climate change and artificial intelligence), presuppose an optimistic attitude.[22] What would interdisciplinary work contribute to the study of climate change if not solutions, or at the very least, the promise of progress towards solvency? In this section, we classify the forms of interdisciplinary optimism that account for the majority of ongoing research and evaluate their efficacy.

From the perspective of the humanities, the most readily available form of interdisciplinary optimism is the deployment of scientific method on subjects formerly reserved for criticism, appreciation, or interpretation. Consilience, E. O. Wilson's familiar argument that all phenomena derive from and are legible as atomic matter (*Moby-Dick* being nothing more than a composition of atoms composed by a brain composed of atoms), is perhaps the example par excellence.[23] According to this model, scientific explanation can in fact explain the entirety of human experience, and the humanities have little to give the sciences except their objects of study as a golden calf. Despite this asymmetry, a great deal of interdisciplinary research on the human mind, literature, and the environment defaults to some version of consilience. Consider, for example, the so-called literary

Darwinism, which attempts to understand novels and poems as indexes of evolutionary development in the human species, and "neuroaesthetics," which strives for a biological and brain-based theory of art.[24] Conversely, the extension of humanities methods to topics usually reserved for scientific research—almost always, unsurprisingly, by humanists—rarely leads to claims meaningful or intelligible to scientists. Jake Kosek's *Understories: The Political Life of Forests in New Mexico,* an acclaimed anthropology of resource extraction and environmental politics, includes a lyrical commentary on nuclear testing that adheres to this logic. "Living next to a deeply secretive, historically deceptive nuclear research complex that produces a highly volatile, mobile, odorless, tasteless, invisible substance that is unimaginably enduring and deadly in its toxicity blurs the traditional boundaries between material and imaginary," Kosek writes in a lucid theory of how nuclear testing alters the assumed parameters of nature itself.[25] Kosek goes on to explore the "haunting" persistence of radiation in different objects and sites. One could imagine a scientist admiring such a powerful and creative account of nuclear testing's impact on human and nonhuman communities. But it is difficult to imagine a scientist recruiting Kosek's work even indirectly, even in a grant-writing context.[26] The transmission back and forth across the divide might be smooth, but the use-value of the optimistic "product" is dubious. The unwanted foray of one discipline into another can actively work against the production of shared knowledge.

Without passing judgment on the value of these unlimited cross-applications, the desire for science to function normatively frequently underwrites these limitless and reductive types of optimism and substitutes for a scientific payoff. Whether imagined as either ideology critique or teleology, a normative science is usually figured as merging with or adopting the political imperatives of the humanities.[27] How might the sciences change the objective reality that they explain? For the better one would presume? The politics at play, indebted to Marx and Hegel, are almost always liberating and oppositional, if not outright leftist. Today, in the American public sphere, the query of how Kosek might speak to scientists is answered with the figure of the scientist-activist, who defends the pursuit of science as simultaneously outside the political, contingent on a particular politics, and normative. The possibility remains that a type of interdisciplinary activism, deriving from either consilience or its converse, could solve the paradox of smooth transmission without utility. "Sci-activism" attempts to preserve climate science, but an activism without such a programmatic vision might be pessimistic. In the absence of laboratories where poems can be pipetted and centrifuged, and in a more limited sense than consilience or its converse, researchers have adopted experimental techniques and computational models to test hypotheses about topics as diverse as the genres of novels and the speeches of American presidents.[28] Known in the humanities at large as the "digital humanities" (and occasionally in literary studies as "distant reading"), constructing digital archives and large text corpuses and applying statistical analysis to the data involves both, as Steven E. Jones notes, "using computers to research literature or art or history" and employing "the methods, insights, and research questions of the humanities to the study of computing and digital media."[29] This immense and complex field proves difficult to characterize as a whole; however, its basic premises are that cultural artifacts are susceptible to algorithmic processing and therefore falsifiable scientific explanation, and that such explanation yields knowledge valuable to humanists and, maybe, scientists. Digital humanities approaches to the study of nuclear testing regimes and oceanography abound. The National Endowment for the Humanities' Digital Humanities Start-Up Grant for "Re-Framing the Online Video Archive: A Prototype Interface for America's Nuclear Test Films" is just one example. The project statement expresses a primary interest in "the visual representation of nuclear weaponry as a morally acceptable tool of the State. Our interest is in learning how photographs and films of human operators at work on such systems have helped citizens address, dismiss or ignore the moral questions surrounding atomic warfare."[30] To accomplish these goals, the investigators archived test films in digital formats using metadata like tags on individual videos and documentary materials keyed to those tags. The digital humanities are optimistic insofar as they apply quantitative, empirical methods to artifacts from the history of science in order to draw confident, scientifically verifiable conclusions about the world. The optimistic classification is not to disparage the digital humanities' validity or value. Rather, we hope simply to recognize the subsuming tide of optimism, with its untested premises, in the development of these projects.

"Re-Framing the Online Video Archive" also represents two ways that a scientific perspective might conceive and use optimistic interdisciplinary research: using the humanities to improve communication of research to the public and adapting to the insights of the sociology or history of science to optimize methods. The humanities as helper, sciences as hamburger. Both pedagogical and progressive modes beg the question—what if a smooth transmission from sciences to humanities—here, designing museum exhibits or websites or studying the internal processes and histories of scientific fields in order to improve science—is not so smooth or does not in fact educate or improve? Of course, optimistic interdisciplinary research can be smooth and useful. But a default to the optimistic forecloses entire fields of interdisciplinary collaboration that do not beg the question, but instead experiment with low permeability between the disciplines or shift attention away from instrumental research outcomes.

Interdisciplinary Pessimism

JB: How interesting, provocative, or stimulating of a node is either surf punk or bomb radiocarbon in organizing this project?

FP: I've found it stimulating but difficult.

JB: What was difficult about it?

FP: With bomb radiocarbon, it's trying to think of it outside my usual terms, which is a tool to understand physical processes on Earth. And instead, we're using it as a tool to understand how you organize ways of thinking or ways of studying things. I'm thinking of bomb radiocarbon in a way that I never think about it. I would have never thought about it otherwise except as a hammer with which to hit the nail of, "how fast does the ocean mix?"

JB: I'm surprised by how much this project has improved the quality of my thinking. Once you create a method and commit to it, you have to interrogate whether or not you're being true to your protocols, and if not, whether that means the method is fundamentally flawed.

Whereas interdisciplinary optimism is infrequently collaborative—
it is difficult to find examples of work in which physical scientists
work directly with humanists under such parameters—the starting
point for interdisciplinary pessimism is collaboration without the
promise or prospect of mutual understanding or utility. Ironically,
promoting open-ended and equal exchange between distant disciplines requires no expectation that the process will generate positive
knowledge. And despite optimism's stranglehold, open-ended and
equal exchange is ongoing in the status quo as a mode that almost
never receives explicit institutional support. With funding limited
for even the most obviously justifiable interdisciplinary projects, optimism beats pessimism in the scrabble for grant dollars. Further,
when the two modes are brought into proximity, optimistic interdisciplinarity tends to coopt pessimistic by turning its internal processes toward predetermined ends. Ideally, a more moderate version
of pessimistic interdisciplinarity might coexist uneasily with the optimistic, agitating and antagonizing.

One modest kind of interdisciplinary pessimism is informal conversation. Academic caricatures notwithstanding, the structure of
the contemporary university is inhospitable to chitchat. Science
and humanities departments are often located at opposite ends of
a campus. Few venues exist to promote free dialogue, except where
project-oriented funding schemes exist. One maligned proposal for
the future of higher education goes all in on the latter idea. A division
of the university by subject or problem area; for instance, a "Water
Program" that would study the entire range of earth science, infrastructural, and cultural systems involving, well, water, has been criticized as the next logical step in the university's corporatization.[31]
Indeed, a "Water Program" seems to embody everything bad about
interdisciplinary optimism, including its bias towards instrumental-
ity, "real world" solutions, and profit. Managerial, business-sensitive
planning, however effective it may be, fosters a vision of the world as
reducible to case studies. Yet one could imagine semi-formal innova-
tions in a university that would replicate the "Water Department"
without an insistence on "problem solving." Breaking down barriers in the built environment—creating shared office spaces, lounges,
classrooms, and even housing—might foster friendship without an

emphasis on linear collaboration in which outcomes are predetermined and limited.

At its heart, informal talk is nonconclusive. There is no "point" to conversing about the weather, sports, movies, music, families, or food—no purpose beyond encouraging the amicability key to the corporate workplace, a cynic might suggest. Routine friendliness can, however, reflect a genuine concern for others and lead to unexpected connections. Shared interests and deep bonds develop in ordinary interaction, even if work eventually transcends the water cooler. Solidarity, mapped around a set of habits of mind and intellection, not politics or material enrichment or institutional survival, offers nonlinear paths to nondeterministic knowledge creation.

An object like bomb radiocarbon, legible in both the sciences and the humanities, does not *prima facie* demand any sort of interdisciplinary work. Achieving better understandings of nuclear testing regimes, their history, politics, cultural representations, and consequences for earth science does not require humanists to adopt scientific methods, or vice versa, nor does it require humanists and scientists to collaborate in instrumental or noninstrumental ways. Yet interdisciplinary pessimism, in the form of nonconclusive talk, could uncover previously unrecognized "problem spaces" or methods for thinking about bomb radiocarbon. What follows nonconclusive small talk can seem nonsensical, absurd, or pointless, according to the accounting of the corporate university. By talking (writing) and thinking about the genre of "surf punk," we hope to reveal some shriveled fruits of interdisciplinary pessimism. Rather than conducting a reading of surf punk that analyzes its aesthetic forms or ideological content, that purely contextualizes surf punk in the history of nuclear testing, or that assesses its contributions to scientific research on radioisotopes, we engage in a process of conversation, represented here through transcriptions. The following thoughts about surf punk, generated through nonconclusive talk, ratify the bounds of our respective disciplines while deriving from no parent methodology. Talking about the surf punk–bomb radiocarbon nexus in nonconclusive ways opens a greater range of possible outcomes than the optimistic approaches outlined above. Although these outcomes may seem ineffectual or diluted in comparison to actionable strategies or material knowledge, they are ultimately better adapted to the environmental timescales of nuclear testing.

FP: Why did you choose surf punk? It was on our syllabus for a class on "Oceans in Science and Literature" that we designed a year and a half ago, before we started working on this essay. I remember thinking then, "I don't even know what surf punk is, but it sounds like something that exists."

JB: That's what I'm interested in, the "I don't know what this is" phenomenon. Most people know what surf music is and what punk is. Surf punk sounds like it should exist and does exist, but it's hard to say what it actually is, except that surf punk takes the practices and aesthetics of surf culture to the darkest, most abject places. If you watch a movie like *The Endless Summer* from 1966, a movie where people go to surf beautiful beaches, or if you listen to the Beach Boys, it's all about how surfing makes you feel good vibrations and live a fulfilling life. Punk music is antithetical to that worldview. I think what punk rockers see in surf music is a freedom sympathetic to the punk lifestyle. Surfing embodies a way of being free in "nature," mediated through a hypercommercialized culture, and surf punk is an appealing distortion of that relation to the ocean.

FP: So do you view surf punk as explicitly rejecting the capitalist overtures in surf music?

JB: Surf punk carves out a protected space in the feedback loop of commercializing rock. The genre is, in a certain sense, not always that great. Agent Orange's cover of "Miserlou," right? It's not as good as Dick Dale's version. Agent Orange puts a punk rock beat over the surf guitar riff, and now we have surf punk? It doesn't always come off successfully. It's hard to commercialize because it can slide into a bad kind of bad.

FP: Its failure is part of its virtue.

JB: Absolutely.

FP: That sounds suspiciously like what we're talking about with interdisciplinary pessimism. Not because it's unsuccessful, but because it has the option to be unsuccessful.

Surf punk is a weak genre—a "classifying principle" that bleeds around the edges.[32] A few music writers invoke it to describe movements within the broader surf and punk rock genres; the Wikipedia page on "Surf Music" gives it a short subhead; there is even a semiobscure Malibu band, founded in 1976, called the Surf Punks.[33] But no heuristic separates surf from punk from surf punk aside from an ad hoc judgment, "I know it when I hear it." The "it" that consistently arises according to this judgment is defined by an uptake of surf guitar riffs and themes—hanging out at the beach, watersports and adventures, fun in the sun, and California girls and boys—into a noise-centric and nihilistic punk frame. Nostalgia for the 1950s and 1960s rendered dark and distorted, the naivety and optimism of those periods refracted through a post-Vietnam, post-Nixon lens. The *Encyclopedia of Music in the 20th Century* calls surf punk an '80s "revival of the original surf music," citing the band Forgotten Rebels's record "Surfin' on Heroin," and it seems fair to grant the genre at least this one solid contour.[34] Other bands clustered around this moment include JFA (established in 1981 and more closely associated with skate punk), which mashed surf classic "Pipeline" by the Chantays with "Police Truck" by the Dead Kennedys to cut "Pipe Truck"; Agent Orange (founded in 1979), which covered among other surf standards "Miserlou," a traditional Mediterranean song made surf famous by Dick Dale; and the aforementioned Surf Punks, authors of original surf punk compositions like "Surfer's Nitemare," "Beer Can Beach," and "Shark Attack."[35] These surf punk bands may be minor figures in the longer histories of surf and punk rock, but the surf punk genre influenced the music of several more popular groups like the Ramones and the Clash. Mainstream surf punk tracks—in particular, the Clash's "Charlie Don't Surf"—offer a unique if unproductive reflection on nuclear testing's release of bomb radiocarbon into the oceans.

Introduced on the fifth side of their triple album *Sandinista!* (1980), "Charlie Don't Surf" responds directly to Francis Ford Coppola's 1979 film *Apocalypse Now,* an adaptation of Joseph Conrad's novel *Heart of Darkness* to the Vietnam era. Martin Sheen's Marlow-*cum*-Captain Benjamin L. Willard must assassinate Marlon Brando's Kurz, rewritten as a Green Beret gone rogue in the jungles of Cambodia. *Apocalypse Now* has inspired decades of debate. Is the movie pro-war, antiwar, or ambivalent? Viet Than Nguyen's recent novel *The Sympathizer* bypasses this dilemma to highlight Coppola's racist reduction of Vietnamese people to animate props, but we might locate the crux of the pro- versus antiwar debate in the film's strange fixation with surfing. Willard's team includes Lance, "a famous surfer from the beaches south of LA" who water skis to the Rolling Stone's "(I Can't Get No) Satisfaction," and it turns out that Robert Duvall's Lieutenant Kilgore, a maniacal commander who loves nothing more than battle, also appreciates a good wave. In the middle of combat, Kilgore offers his canteen to a wounded enemy soldier, but stops his obscene performance of humanitarian mercy to greet Lance and tell him, "we do a lot of surfing around here." Kilgore chooses the insertion point for Willard's boat because of its surfing prospects, and when one of his soldiers objects that the Viet Cong control the neighboring village, Kilgore rebuts, "Charlie don't surf!" The line, clearly the Clash's inspiration, is delivered by a character both mythologized and reviled in the film. Because the Vietnamese, homogenized and denigrated through a racist slur, do not surf (or participate in American capitalism and its culture industries), they cannot pose real resistance to Kilgore's real objective, riding a tube. By directing the mission around the perverse desire to surf, Kilgore suffuses an uncomplicated, nostalgic vision of American masculinity, the surfer, with racial violence. Surfing is compromised and redeployed for ends other than good vibrations: an ironic performance of nonchalance in the face of death. "Charlie don't surf" comes to represent the governing order of American warfare, a disregard for the lives of civilians on the grounds of their non-Westernness.

FP: I would never willingly listen to surf punk music.

JP: Even "Charlie Don't Surf?"

FP: I think that "Charlie Don't Surf" was actually my least favorite of the surf punk music.

In "Charlie Don't Surf," the Clash adopts the riffs and themes of surf punk to situate the genre in the immediate context of Cold War militarism. Rather than Kilgore's "Charlie don't surf," the Clash sing, "Charlie don't surf and we think he should." The song both ventriloquizes Kilgore, confirming his racist position as true, and expresses a normative, counterfactual claim. The line might be read as a call for targets of Cold War imperialism to Westernize in order to avoid destruction or a plea for resistance. But regardless of capitulation or revolution, the end is unavoidable: the apocalyptic terminus of the geopolitical order. Verses like, "everybody wants to rule the world / . . . Satellites will make space burn," rehearse Kilgore's logic and point to planetary cataclysm. The surf punk genre thus becomes the lexicon through which the unspecified apocalypse of "apocalypse now" is enunciated and mystified. If surfing in *Apocalypse Now* reflects the illogic of the Vietnam War and celebrates its Nietzschean bravado, the Clash translate the film's figure of the surfer into a punk frame to spectate on the Earth's inevitable annihilation. Surf punk's representation of wartime surfing is doubly mediated through this system, extracted through retrospective media like *Apocalypse Now* and the war-making apparatus of the West. Under such conditions, the critical power of punking surf slips into celebration or verification. In this way, "Charlie Don't Surf" might simultaneously salivate over and denounce the Cold War's abhorrent violence.

FP: I had some unverifiable, unknowledgeable opinions about surf punk.

JB: Like what?

FP: Surf punk existed within the margins of two other things without existing on its own at any point. It pushed back without actively pushing back. It advocated for something that it didn't actually do. I see that as a passive counterculture.

As modeled in "Charlie Don't Surf," the surf punk genre holds an indeterminate relationship to the broad context of the nuclear testing regime. Insofar as the 1963 Limited Test Ban Treaty predates the emergence of surf punk by two decades, "Charlie Don't Surf" is necessarily retrospective, like *Apocalypse Now*. In the case of bomb radiocarbon, however, retrospection might serve to analyze ongoing and persistent phenomena, for example, Cold War nuclear fallout localized to a past event (testing) or the specter of nuclear war. The contextual relationship of the Vietnam War to *Apocalypse Now*, structured around the film's localization of the war to the past and refusal to witness its persistent environmental and bodily effects, does not generalize to the relationship between the Vietnam War or the Cold

War and "Charlie Don't Surf" or surf punk. The advantage of surf punk, as a generic mode of representing the violence of war, seems to be its capability to register the expansive and nonhuman timescales over which acts of war like nuclear testing or the invasion of Vietnam persist.

JB: One problem with interdisciplinary optimism is that it's conservative. Interdisciplinary optimism—

FP: Wants to preserve.

JB: To live forever on 1999 Earth, or something like that.

FP: And it can't properly adapt to timescales that we can't imagine. Which is where these arguments about radioactivity come into play, right? People have claimed that mitigation plans for radioactive contamination are nonsense because they play out over such long timescales—

JB: They can't even imagine.

The nexus of surf punk and bomb radiocarbon could only be described as partially visible, indirect, or oblique. Just as bomb radiocarbon indexes the Earth system's response to greenhouse forcing, surf punk indexes scales of radioactive contamination and climate apocalypse. That is, surf punk marks the entanglement of nuclear testing, its underlying motives and intended effects, and anthropogenic shifts in climate systems. In this capacity, surf punk is itself a kind of countercultural context to atmospheric testing's continuous present and unforeseeable future, not to mention its applications in oceanographic research about our changing climate. As opposed to theories of art and cultural production that posit a critical or resistant axis latent in the text, or countercultures that mobilize activists, protesters, and artists, surf punk is passive. "Charlie Don't Surf" does not criticize *Apocalypse Now* so much as bring it into contact with the environmental traumas of the Cold War and their enduring, suprahuman properties. Faced with a phenomenon that exceeds the scale of human civilization, biology, or experiential comprehension, a passive counterculture like surf punk is oppositional and inactive.[36]

Surf punk's technique is to direct attention, not to determine conclusions, if any. An interdisciplinary conversation about surf punk might therefore devolve, tautologically, to recursive conversation. Would a mere impetus to talk about talking be enough to count as interdisciplinary work of some fashion, even or especially if it led to no concrete hypotheses or interpretations? Perhaps talking collaboratively about surf punk's contextualizing features might suggest new structures for organizing interdisciplinary climate studies that would conform to surf punk's mash-up paradigm. According to this thesis, to uptake surf into punk while preserving their respective particularities might be to uptake one discipline into another without instrumental, purposive, or conclusive goals.

FP: Why would anyone spend their time doing what we've been doing?

JB: That question is trapped in a way of thinking that prescribes outcomes.

FP: Your job is dependent on generating outcomes for your university, like getting a grant, right?

Analogical reasoning risks a kind of disciplinary nihilism, in which one discipline seeks to mutilate, distort, and obliterate another. One alternative to this negativity is the pursuit of creativity. Taking cues from the Clash might not be the best way to research climate change, but it might not be the worst. Making art, or more generally, participating in creative projects that preserve disciplinary boundaries, could represent one appealing form of pessimistic climate studies. The resulting artworks and their relative degrees of political efficacy or knowledge-generation are irrelevant. A background state, analogous to surf punk's passive counterculture, out of which the artwork emerges is a possible outcome of interdisciplinary pessimism.

The injunction to abandon interdisciplinary optimism and start bands or artists' collectives or talking about stuff is vulnerable to objection on the grounds of absurdity. But this essay's advocacy is not a dadaism or nihilism, as one misinterpretation of surf punk might suggest. Kurtz's final warning, that "horror has a face and you must make a friend of horror. Horror and moral terror are your friends. If they are not, they are enemies to be feared," is not the necessary correlate of pessimism. Surf punk's insight is that objects of study like bomb radiocarbon or climate change invite new disciplinary protocols that do not assume either the smooth transference of knowledge and method across disciplines or the production of usable knowledge and practical solutions. Nuclear testing, the Cold War, and Vietnam provoke in surf punk an apprehension of what cannot be apprehended directly: radioactivity and anthropogenic warming emanating from the military-industrial complexes of the "superpowers." Interdisciplinary pessimism ought to aim its gaze at these catastrophes as they unfold on timescales imperceptible to human life.

FP: I don't know, is that also grounded in the late-nineties upper middle-class, white American values?

JB: Probably, but my way of getting around that is to imagine a counterfactual in which you're a colonist on an alien planet.

FP: Like a desert planet where it rains once every thousand years?

JB: Or even more outside the realm of human imagination. We live on a planet that orbits a red dwarf, where instead of chlorophyllic photosynthesis, photosynthesis happens in the infrared spectrum. Everything is black. It's hard to imagine the human organism not going insane under those conditions. Right?

FP: I would agree.

JB: There seems to be some kind of problem with preserving the category of the human and also imagining a solution to climate change.

FP: We have no idea what the future might hold, what the outcomes should be, or what timescales they evolve over. The advantage of interdisciplinary pessimism is that it encompasses unimaginable things as they come up.

NOTES

1. These conversations are transcribed from recordings made by Jason and Frankie in May 2017.

2. Willard F. Libby, "Atmospheric Helium Three and Radiocarbon from Cosmic Radiation," *Physical Review* 69 (1946): 671–72.

3. Hiroyuki Kitagawa and Johannes van der Plicht, "Atmospheric Radiocarbon Calibration to 45,000 yr B.P.: Late Glacial Fluctuations and Cosmogenic Isotope Production," *Science* 279 (1998): 1187–89.

4. Reidar Nydal, "Increase in Radiocarbon from the Most Recent Series of Thermonuclear Tests," *Nature* 200 (1963): 212.

5. U.S. Department of State, The Treaty Banning Nuclear Weapon Tests in the Atmosphere, in Outer Space and Under Water with U.S.S.R. and U.K., August 5, 1963, 14 U.S.T. 1313, T.I.A.S. No. 5433.

6. Quan Hua, Mike Barbetti, and Andrzej Rakowski, "Atmospheric Radiocarbon for the Period 1950–2010," *Radiocarbon* 55 (2013): 2059–72.

7. Reidar Nydal and Knut Lövseth, "Tracing Bomb ^{14}C in the Atmosphere 1962–1980," *Journal of Geophysical Research* 88 (1983): 3621.

8. Ingeborg Levin and Vago Hesshaimer, "Radiocarbon—A Unique Tracer of Global Carbon Cycle Dynamics," *Radiocarbon* 42 (2000): 69.

9. Minze Stuiver, "^{14}C Distribution in the Atlantic Ocean," *Journal of Geophysical Research* 85 (1980): 2711–18.

10. Ulrich Siegenthaler and Fortunat Joos, "Use of a Simple Model for Studying Oceanic Tracer Distributions and the Global Carbon Cycle," *Tellus* 44B (1992): 186–207.

11. See Dipesh Chakrabarty, "The Climate of History: Four Theses," *Critical Inquiry* 35 (Winter 2009): 220–22; and Jana Silliman et al., "Climate Emergencies Do Not Justify Engineering the Climate," *Nature Climate Change* 5 (2015): 290–92.

12. See Matthew Collins et al., "Long-term Climate Change: Projections, Commitments, and Irreversibility," in *Climate Change 2013: The Physical Science Basis*, ed. T. F. Stocker et al. (Cambridge: Cambridge University Press, 2013), 1029–106.

13. See, for example, Rafael Reuveny, "Climate Change–Induced Migration and Violent Conflict," *Political Geography* 26 (August 2007): 656–73; and Céline Bellard et al., "Impacts of Climate Change on the Future of Biodiversity," *Ecology Letters* 15 (2012): 365–77.

14. For an example of a proposed technological fix to combine with mitigation efforts, see T. M. L. Wigley, "A Combined Mitigation/Geoengineering Approach to Climate Stabilization," *Science* 314 (October 2006): 452.

15. Susan Solomon, Gian-Kasper Plattner, Reto Knutti, and Pierre Friedlingstein, "Irreversible Climate Change Due to Carbon Dioxide Emissions," *PNAS* 106 (2009): 1704.

16. See, for example, Hans-Martin Füssel, "Vulnerability: A Generally Applicable Conceptual Framework for Climate Change Research," *Global Environmental Change* 17 (May 2007): 155–67. One recent development in this area is the start of coalitions to preserve research resources in the face of political uncertainty.

17. Colin N. Waters et al., "Can Nuclear Weapons Fallout Mark the Beginning of the Anthropocene Epoch," *Bulletin of the Atomic Scientists* 71 (2015): 46–57.

18. To a certain extent our technique comes to resemble Lee Edelman's vision for "the impossible project of a queer oppositionality," which refuses "history as linear narrative (the poor man's teleology) in which meaning succeeds in revealing itself—*as itself*—through time." See Lee Edelman, *No Future* (Durham, N.C.: Duke University Press, 2004), 4. What would an interdisciplinary method look like with no future?

19. See Jonathan Kramnick and Anahid Nersessian, "Form and Explanation,"

Critical Inquiry 43 (Spring 2017): 650–69; Jonathan Kramnick, "The Interdisciplinary Fallacy," *Representations 140* (Fall 2017): 68.

20. This short statement about falsifiability derives almost entirely from the work of Karl Popper. Whereas the sciences falsify, the humanities prove—that is, the emphasis on argument in the humanities is a sign that what humanists do is methodologically distinct from scientific research. See Karl Popper, *The Logic of Scientific Discovery* (London: Taylor and Francis, 2005), 18.

21. C. P. Snow, *The Two Cultures and the Scientific Revolution* (New York: Cambridge University Press, 1961), 4.

22. See Carole L. Palmer and Laura J. Neumann, "The Information Work of Interdisciplinary Humanities Scholars: Exploration and Translation," *The Library Quarterly* 72 (January 2002): 85–117; *Practicing Interdisciplinarity*, ed. Peter Weingart and Nico Stehr (Toronto: University of Toronto Press, 2000); Robert Frodeman, ed., *The Oxford Handbook of Interdisciplinarity* (Oxford: Oxford University Press, 2010); Andrew Barry, Georgina Born, and Gisa Weszkalnys, "Logics of Interdisciplinarity," *Economy and Society* 37 (2008): 36–38, 41–42; and Moti Nissani, "Ten Cheers for Interdisciplinarity: The Case for Interdisciplinary Knowledge and Research," *The Social Science Journal* 34 (1997): 201–16.

23. Edward O. Wilson, *Consilience: The Unity of Knowledge* (New York: Vintage, 1999). See also *Darwin's Bridge: Uniting the Humanities and Sciences*, ed. Joseph Carroll, Dan P. McAdams, and Edward O. Wilson (New York: Oxford University Press, 2016), and Edward Slingerland, *What Science Offers the Humanities: Integrating Body and Culture* (Cambridge: Cambridge University Press, 2008).

24. See Joseph Carroll, *Literary Darwinism: Evolution, Human Nature, and Literature* (New York: Routledge, 2012), and Anjan Chatterjee, "Neuroaesthetics: A Coming of Age Story," *Journal of Cognitive Neuroscience* 23 (January 2011): 53–62.

25. Jake Kosek, *Understories: The Political Life of Forests in Northern New Mexico* (Durham, N.C.: Duke University Press, 2006), 258–59.

26. We are open to the idea that programs like the National Science Foundation's Science, Technology, and Society grants can produce interesting pessimistic interdisciplinary work.

27. Roy Bhaskar's "critical realism" offers one scheme for this merger. See Roy Bhaskar, *Scientific Realism and Human Emancipation* (London: Routledge, 2009).

28. See Sarah Allison et al., "Quantitative Formalism: An Experiment," *Stanford Literary Lab*, January 15, 2011: 3–6, and Dan Faltesek, "Obama's Sixth Annual Address: Image, Affordance, Flow," *Digital Humanities Quarterly* 10 (2016), http://www.digitalhumanities.org/dhq/vol/10/4/000280/000280.html.

29. Steven E. Jones, *The Emergence of the Digital Humanities* (New York: Routledge, 2014), 6.

30. Kevin Hamilton and Ned O'Gorman, "Re-Framing the Online Video

Archive: A Prototype Interface for America's Nuclear Test Films," *NEH5*, http://illinoisneh5o.omeka.net/exhibits/show/pastfive/reframing, accessed May 18, 2017.

31. The proposal in question emerges in an op-ed by Mark C. Taylor, "End the University as We Know It," *New York Times*, April 26, 2009.

32. Wai Chee Dimock, "Critical Response I: Low Epic," *Critical Inquiry* 39 (Spring 2013): 623.

33. See Konstantin Butz, *Grinding California: Culture and Corporeality in American Skate Punk* (New York: Columbia University Press, 2012); Marc Spitz and Brendan Mullen, *We Got the Neutron Bomb: The Untold Story of L.A. Punk* (New York: Random House, 2001); Colin J. Campbell, "'Three-Minute Access': Fugazi's Negative Aesthetic," in *Adorno and the Need in Thinking*, ed. Donald A. Burke, Colin J. Campbell, Kathy Kiloh, Michael K. Palamarek, and Jonathan Short, 291 (Toronto: University of Toronto Press, 2007); Timothy J. Cooley, *Surfing about Music* (Berkeley: University of California Press, 2014); and Dewar MacLeod, *Kids of the Black Hole: Punk Rock in Postsuburban California* (Norman: University of Oklahoma Press, 2010), 121.

34. Judi Gerber, "Surf Music," in *Encyclopedia of Music in the 20th Century*, ed. Lee Stacey and Lol Henderson (New York: Routledge, 1999), 619.

35. Bill Brown claims that JFA looked "uncomfortable, even a little guilty" during performances of their "surf/punk" repertoire. Bill Brown, *You Should've Heard What I Just Seen: Collected Newspaper Articles, 1981–1984* (Cincinnati: Colossal Books, 2010), 106.

36. Rei Terada theorizes oppositional inactivity as the Romantic "impasse" in her essay "Looking at the Stars Forever," *Studies in Romanticism* 50 (2011): 280. According to her account, an "impasse is a kind of barricade to create space for a world in which futility can no longer be a reason for not doing something."

CHAPTER 2

Earth's Changing Climate
A Deep-Time Geoscience Perspective
JANE E. DMOCHOWSKI AND DAVID A. D. EVANS

The Earth is more than four billion years old, and animal life has occupied only the past twelve percent of the planet's history. Our hominid ancestors arrived so late on the scene that, if geological time were compressed into a calendar year, the well-known hominid fossil Lucy would be walking across the African savannah just six hours before midnight on New Year's Eve. All ten thousand years (or so) of human civilization would compress into the final minute of the year, and with industrialization occurring just one second before the stroke of midnight, we find our species exploding in population, dominating nearly every habitat and ecosystem, and mobilizing nearly all of the chemical periodic table to satisfy our hunger for technological development. In short, we are accelerating headlong into the Anthropocene New Year.

If we could take an outsider's perspective on this explosive development, we might wish to pause and reflect upon whence we have come, and whither we are headed. Geology affords that synoptic vantage, gained through the methodical establishment of a global time-scale that is calibrated in "absolute" time by our ability to date rocks using radioactivity. Carbon dating is the method best known to the general public, but longer-lived radioactive isotope decay systems are used extensively to corroborate and refine the geologic time-scale: uranium to lead, potassium to argon, samarium to neodymium, among others. In just the past century, the science of dating Earth history, or geochronology, has progressed to such a level among the world's best laboratories that the precision of an age can now be refined to about 0.01 percent. As an example, two volcanic ash beds in sedimentary strata from Colorado are found to bracket the extinction horizon for the demise of dinosaurs. Minute crystals of the mineral zircon extracted from the lower, older ash bed are dated by uranium–lead decay at 66.082 +/-0.022 million years ago; and the younger, upper ash bed is dated by the same method at 66.019 +/-0.024 million years ago.[1] We can celebrate the fact that our state-of-the-art measurements are able to achieve such unprecedented precision, but we also must realize that in absolute terms, uncertainty around the age of dinosaur extinction spans more than 60,000 years—still far too great an interval to get at some of the key questions about why and how it occurred (i.e. how much of it was caused by the Deccan volcanic eruptions in India, versus the meteor impact in Mexico). Such is a testament to the enormity of the geologic record and the challenges we face in understanding events in so-called "deep time."

Despite the limitations of establishing absolute geochronometers, chemical clues in sedimentary rocks indicate unambiguously that climate has changed dramatically in the geologic past. The climate variations are due to changes in solar radiation received at the top of Earth's atmosphere and concentrations of greenhouse gases within the atmosphere. The atmospheric warming effect is on the order of tens of degrees Celsius, and liquid water would surely not exist on the surface of our planet without its greenhouse envelope.[2] In 1863, John Tyndall famously described water vapor, the first recognized greenhouse gas, as providing "a blanket more necessary to the vegetable life of England than clothing is to man."[3] Modern research identifies a host of different molecules that have the greenhouse-gas characteristics of transmitting shorter-wavelength solar radiation to the planet's surface but absorbing the longer-wavelength outgoing radiation emitted by the Earth: water vapor, methane, carbon dioxide (CO_2), and others. Among these, increasing water vapor concentration readily leads to condensation and thus does not allow a substantial runaway effect of warming. Methane quickly oxidizes (to water and CO_2) so by itself only has a local or short-term significant greenhouse effect. Carbon dioxide, currently at ~415 parts per million (ppm) concentration (above a ~280 ppm pre-industrial background level), is far below its saturation pressure and has the capacity to accumulate substantially; it is also stable in our oxidized atmosphere and is a direct byproduct of both natural and anthropogenic hydrocarbon combustion. For these reasons, CO_2 is the greenhouse gas of greatest importance to long-term climatic change on Earth.

The geologic record of past climate changes shows remarkable variations. The continuous presence of liquid water on the Earth's surface, as indicated by water-lain sediments now preserved as sedimentary rocks, divulges a remarkable stabilizing feedback in spite of the sun growing more luminous through the last four billion years. The feedback involves temperature-modulated rates of silicate rock weathering, which removes atmospheric CO_2 and ultimately deposits the carbon in sedimentary rocks, such that hotter climate leads to more CO_2 removal, and cooler climate leads to less removal[4]— processes that tend to stall any developing trend toward either warming or cooling. Departures from this stable long-term state include the catastrophic "Snowball Earth" ice ages that engulfed the planet in ice for millions of years at a time, prior to the evolution of animal life.[5] We still don't know exactly why the silicate-weathering feedback of climate stability failed on those occasions, but they are limited to eras prior to the evolution of complex animal and plant life. For the past 500 million years, when complex life has teemed across the Earth's surface, ice ages have waxed and waned but remained moderate in areal extent; it's possible that numerous biologically mediated processes in an increasingly complex carbon cycle have helped avoid runaway climate catastrophe during the more recent chapters of Earth history.[6] Within that late time interval, global climate has oscillated on roughly a 300–400 million year quasi-cycle, which is generally thought to correspond to the assembly and dispersal phases of the Pangea supercontinent and the southward growth of the Eurasian landmass—all connected to climate via enhanced or reduced rates of carbon fluxes between the solid Earth and the atmosphere, for example volcanic outgassing of CO_2.[7]

The latest glacial interval spans the past 35 million years in Antarctica and the past 2–3 million years in the northern high latitudes. Detailed borehole core records from ocean sediments and ice caps show that the latter epoch's climate has been paced by Earth's orbital cycles, for example the axial obliquity (tilt) and the orbital eccentricity around the sun (noncircularity, modulating an additional precession of the equinoxes in celestial coordinates). These variations, called Milankovitch cycles, operate at periodicities between 20,000 and more than 100,000 years. If operating alone, they would currently be sending us slowly toward the next ice age.[8] As seen from these examples, many diverse processes control climate change on Earth, at a range of timescales. Deniers of anthropogenic influence on the climate will emphasize the existence of these past changes to state that climate variability is purely a natural phenomenon, but such logic is specious: human influence on the climate through industrial atmospheric carbon emissions should be viewed as an enhancement of carbon cycle fluxes that are well known to warm the planet's surface.

One paleoclimatic event of particular interest to possible future global warming scenarios is known as the Paleocene-Eocene Thermal Maximum (PETM). That event occurred about 55 million years ago, as an abrupt episode of organic carbon released into the atmosphere as either methane or CO_2, with a coincident spike in global temperature and significant ecosystem shifts. The PETM is regarded by climatologists as a possible ancient analog for the current anthropogenic release of about 500 billion tons of carbon into the atmosphere via fossil-fuel combustion.[9] It is the most dramatically rapid prehistoric perturbation to climate and the carbon cycle that we know of, so scientists would naturally want to understand details of how the planet responded and returned to normalcy. Unfortunately, the PETM's antiquity implies temporal precision on the same order as that of the dinosaur extinction, described above. If we want to use insights from the PETM or other deep-time climate perturbations to make predictions about our own current anthropogenic climate experiment, we are only able to estimate the consequences on timescales of about 50,000 years into the future. This is unsatisfactory.

So we turn to theoretical understanding of an anthropogenic greenhouse world. During the late nineteenth century, following Tyndall's pioneering study, physicists strove to quantify the temperature rise accompanying atmospheric greenhouse gas accumulation. Svante Arrhenius discovered a nonlinear relationship whereby each doubling of atmospheric CO_2 yields a fixed change in temperature, now referred to as "climate sensitivity."[10] It is generally thought to be between 1.5 and 4.5 degrees Celsius (°C) per doubling of CO_2.[11] The variation in estimates results from our incomplete understanding of climate feedbacks, including the effects of water vapor, aerosols, ice, and cloud cover. Clouds, for example, both reflect solar radiation back to space and trap long wavelength infrared radiation emitted by Earth's surface. Simulated cloud cover under warming conditions has been shown to increase in area,[12] but also decrease in thickness.[13] The overall net climatic effects are unclear.

News outlets frequently pronounce new highs in our global average atmospheric CO_2 levels, but these in-place measurements, starting with those measured at the Mauna Loa Observatory, only go back to 1958. These data have been augmented in the past 15 years with global CO_2 measurements via satellites. We also have measurements of CO_2 going back 800,000 years from ice cores, but these data are less accurate than modern measurements, with the measurement of CO_2 known to just a few parts per million by volume (ppmv),[14] compared to an accuracy of <0.2 ppm for measurements made at the Mauna Loa Observatory,[15] and their temporal measurements only accurate to a few percent of the true age. Thus, a CO_2 measurement made for ice determined to be 130,000 years old could be as old as 136,000 or as young as 124,000 years old.[16] This relative dearth of greenhouse gas data until very recent times, combined with the sparsity of other climate-relevant data (land-cover, cloud-cover, solar variations, oceanic oscillations), complicates the process of testing our general circulation models (GCMs, also informally known as global climate models). How do scientists know if our future predictions are accurate without waiting for the future to happen? They attempt to predict past changes with their model and determine how accurate those predictions are. But determining how well the model matches observational data requires that we have more complete observational data.

We know that global temperatures have been rising significantly over the last century, along with CO_2 levels. The global mean temperature for 2000–2010 was 0.8 °C warmer than the 1880–1920 mean, with over 0.5 °C of that increase occurring since 1975.[17] GCM simulations incorporating natural phenomena can only partially explain the rise in temperatures over the past century, indicating that much is a result of human activities.[18] Using likely greenhouse gas concentration scenarios and our understanding of Earth's energy balance—equivalent to incoming solar radiation minus the reflected component—our current GCMs can predict future global mean surface temperature rise fairly well. For example, the Coupled Model Intercomparison Project Phase 5 (CMIP5) determines that using the

Representative Concentration Pathway 6 (one of four greenhouse gas concentration scenarios, considered an intermediate scenario, with the "6" indicating a +6 Watt per square meter (W/m^2) radiative forcing in the year 2100 relative to pre-industrial values)[19] has predicted that the mean surface temperature for 2081–2100, compared to the 1986–2005 reference period, will increase by 2.2°C, with a likely range between 1.4 to 3.1°C.[20] To put that number into perspective, for every 1°C increase in global warming it is estimated that there will be a 7–15 percent decrease in yields of several crops in major growing regions, the average intensity of tropical cyclones will increase by 1–4 percent, and extreme precipitation will increase by 3–10 percent.[21] Thus, these models can inform us of the likely consequences of unabated fossil-fuel emissions.

Temperature increases over the last century have already led to societal effects. Due to increased temperatures, sea level rise, and changes in precipitation that can lead to both increased flooding in one place and drought in another, many areas are already experiencing permafrost and sea ice loss, rising sea levels, flooding, and extreme weather events, to name just a few of the effects. In North America, economic and ecological integrity losses, water quality impairment, and human morbidity and mortality are characterized as medium to very high present-day risks.[22]

However, these risks can be at least partially addressed at the personal through global levels. At home we can decrease our water usage, better manage our home cooling and heating, make sure our homes are built to withstand possible storms, live in regions less likely to sustain damage from flooding and wildfires, incorporate sustainable food choices into our diets, and remain informed and willing to adapt. Governments and organizations can provide heat warnings and air conditioning to those in need, help farmers respond to changes in crops, encourage sustainable building practices, and work to encourage communities that can continue to thrive despite changes predicted for the future. Many are beginning to do so. For example, the Philadelphia Water Department has worked to address issues of storm water pollution entering its combined sewer system through its "Green City, Clean Waters" initiative, by increasing permeable surfaces in the city and otherwise "greening" the city's infrastructure.[23] Interagency efforts, such as the California Wildland Fire

Coordinating Group,[24] are collaboratively managing wildfires. And, in Venice, Italy, billions of euros will be spent to complete the MOSE project (MOdulo Sperimentale Elettromeccanico [Experimental Electromechanical Module]), a series of mobile gates designed to protect the Venetian lagoon from flooding.[25] There are even more ambitious efforts proposed, such as floating hotels, golf courses, and even entire cities.[26] And while it is usually proposed with much caution, many individuals and organizations go further and advocate for more research into geoengineering (or climate engineering) schemes, which fall into one of two categories: solar radiation management (e.g., adding aerosols to the stratosphere to increase Earth's albedo) and CO_2 removal (e.g., iron fertilization to increase the CO_2 uptake in the ocean).[27]

To address the underlying cause of climate change, however, we need to decrease our reliance on fossil fuels by decreasing our energy consumption, investing in alternative energy and efficient transportation, decreasing our meat consumption, and carefully mitigating the destructive land-use changes of the past 100-plus years. The Paris Agreement, adopted in December 2015 and signed by 195 countries by April 2017, was a monumental step toward a global effort to mitigate and adapt to climate change that is going into effect in 2020.[28] However, without more ambitious Intended Nationally Determined Contributions (INDCs) than those the signatories have reported, the goal to hold global average temperature to well below 2 °C above pre-industrial levels will not be met.[29]

Deciding on how and when to implement adaptation and mitigation strategies depends on having an improved understanding of the social and natural science affecting each impact. Now is the time to double-down on our efforts to collect and analyze data, and communicate scientific results that are subjected to rigorous peer review. Policy decision-making needs to be based on our best scientific understanding. Continued collaboration between natural and social scientists, humanists, engineers, and policy makers is crucial as Anthropocene climate change is likely to involve continued sea level rise, ocean acidification, increase in waterborne and vector-borne diseases, variation in crop yields, and changing weather patterns.

The effects of climate change may require involuntary displacement of large populations, with hard-to-predict human/economic costs, which will be disproportionately felt by the most disadvantaged individuals and countries. Extreme environmental conditions that put communities, governments, and resources under duress will exacerbate the already difficult tasks of meeting the basic needs of our world's most vulnerable populations. Yet those most to blame live in relatively privileged countries. Just ten countries make up more than 60 percent of total greenhouse gas emissions, with the United States of America contributing 14.1 percent of the total. Meanwhile, seventeen countries, mostly island and African countries, none of which are one of the top-ten contributors, are currently acutely vulnerable to climate change impacts.[30]

Thinking "outside the box," can we explore humanity's long-term survival options beyond Earth? Some would point to interplanetary colonization to increase the likelihood of human survival—to the Moon, Mars, or beyond—but there is no better place for humans in the cosmic neighborhood than our home planet. Earth has ample water, oxygen, a large human population for genetic diversity, and a magnetic field that largely protects us from exposure to cosmic radiation. Earth's distance and daily exposure to the Sun, combined with the climate-moderating effect of the oceans, makes its temperature ideal for human life. Additionally, we have evolved to thrive in Earth's gravity and air pressure (both of which are much lower on Mars). Back on our home planet, adaptation and mitigation seem to be formidable challenges, but they are far more achievable than colonizing moons or other planets. Delaying action will make mitigation more expensive, less effective, and will exclude options that may only be available in a brief window of time.

Returning to the analogy of Earth's calendar year, we arrive at the stroke of midnight on New Year's Eve. Humankind is enjoying an incredible party, fueled by the bountiful harvest of the entire past calendar year of Earth history. There's more of that harvest in the storehouse, but we must decide: will we continue our celebrations as usual, or even at the accelerating pace of "economic growth"? Or, will we embrace renewable energy and move toward complete recycling of materials? The latter strategy may somewhat diminish the exuberance of our midnight celebrations, but with that option we are likely to feel much better in the morning. Geology provides the foundation for, and encapsulates the fate of, industrial society. It has given us the raw materials to develop awe-inspiring technological breakthroughs, and it has also taught us the planetary context of our existence. Now is our chance to use geological insight to design our environmental future.

NOTES

1. William C. Clyde, Jahandar Ramezani, Kirk R. Johnson, Samuel A. Bowring, and Matthew M. Jones, "Direct High-Precision U-Pb Geochronology of the End-Cretaceous Extinction and Calibration of Paleocene Astronomical Timescales," *Earth and Planetary Science Letters* 452 (2016): 272–80.

2. T. Raymond, *Principles of Planetary Climate* (Cambridge: Cambridge University Press, 2010).

3. John Tyndall, "On Radiation through the Earth's Atmosphere," *Philosophical Magazine* (Series 4) 25 (1863): 200–206.

4. James C. G. Walker, P. B. Hays, and J. F. Kasting, "A Negative Feedback Mechanism for the Long-Term Stabilization of Earth's Surface Temperature," *Journal of Geophysical Research* 86, C10 (1981): 9776–82.

5. Paul F. Hoffman, "Pan-glacial—a Third State in the Climate System," *Geology Today* 25 (2009): 107–14.

6. David A. D. Evans, "A Fundamental Precambrian-Phanerozoic Shift in Earth's Glacial Style?" *Tectonophysics* 375 (2003): 353–85.

7. N. Ryan McKenzie, Brian K. Horton, Shannon E. Loomis, Daniel F. Stockli, Noah J. Planavsky, and Cin-Ty A. Lee, "Continental Arc Volcanism as the Principal Driver of Icehouse-Greenhouse Variability," *Science* 352 (2016): 444–47.

8. J. D. Hays, J. Imbrie, and N. J. Shackleton, "Variations in the Earth's Orbit: Pacemaker of the Ice Ages," *Science* 194 (1976): 1121–32.

9. Francesca A. McInerney and Scott L. Wing, "The Paleocene-Eocene Thermal Maximum: A Perturbation of Carbon Cycle, Climate, and Biosphere with Implications for the Future," *Annual Reviews of Earth and Planetary Sciences* 39 (2011): 489–516. Josep G. Canadell, Corinne Le Quéré, Michael R. Raupach, Christopher B. Field, Erik T. Buitenhuis, Philippe Ciais, Thomas J. Conway, Nathan P. Gillett, R. A. Houghton, and Gregg Marland, "Contributions to Accelerating Atmospheric CO2 Growth from Economic Activity, Carbon Intensity, and Efficiency of Natural Sinks," *Proceedings of the National Academy of Sciences, USA* 104 (2007): 18866–70.

10. Svante Arrhenius, "XXXI. On the Influence of Carbonic Acid in the Air upon the Temperature of the Ground," *The London, Edinburgh, and Dublin Philosophical Magazine and Journal of Science* 41, no. 251 (1896): 237–76.

11. Rajendra K. Pachauri, Myles R. Allen, Vicente R. Barros, John Broome, Wolfgang Cramer, Renate Christ, John A. Church et al., "Climate Change 2014:

Synthesis Report." Contribution of Working Groups I, II and III to the fifth assessment report of the Intergovernmental Panel on Climate Change. IPCC, 2014.

12. Minghua Zhang and Christopher Bretherton, "Mechanisms of Low Cloud–Climate Feedback in Idealized Single-Column Simulations with the Community Atmospheric Model, version 3 (CAM3)," *Journal of Climate* 21, no. 18 (2008): 4859–78.

13. Anthony D. Del Genio and Audrey B. Wolf, "The Temperature Dependence of the Liquid Water Path of Low Clouds in the Southern Great Plains," *Journal of Climate* 13, no. 19 (2000): 3465–86.

14. D. Lüthi, "EPICA Dome C Ice Core 800KYr Carbon Dioxide Data," IGBP PAGES/World Data Center for Paleoclimatology Data Contribution Series 55 (2008).

15. Pieter Tans and Kirk Thoning, "How We Measured Background CO2 Levels on Mauna Loa," U.S. Department of Commerce, Nation Oceanic and Atmospheric Administration, Earth System Research Laboratory (2008).

16. Frédéric Parrenin, J-M. Barnola, J. Beer, Thomas Blunier, E. Castellano, J. Chappellaz, G. Dreyfus et al., "The EDC3 Chronology for the EPICA Dome C ice core," *Climate of the Past* 3, no. 3 (2007): 485–97.

17. James Hansen, Reto Ruedy, Makiko Sato, and Ken Lo, "Global Surface Temperature Change," *Reviews of Geophysics* 48, no. 4 (2010): 29.

18. G. C. Hegerl, F. W. Zwiers, P. Braconnot, N. P. Gillett, Y. Luo, J. A. Marengo Orsini, N. Nicholls, J. E. Penner, and P. A. Stott, "Understanding and Attributing Climate Change," in *Climate Change 2007: The Physical Science Basis. Contribution of Working Group I to the Fourth Assessment Report of the Intergovernmental Panel on Climate Change*, ed. S. Solomon, D. Qin, M. Manning, Z. Chen, M. Marquis, K. B. Averyt, M. Tignor and H. L. Miller, 663–746 (Cambridge: Cambridge University Press, 2007).

19. Karl E. Taylor, Ronald J. Stouffer, and Gerald A. Meehl. "An Overview of CMIP5 and the Experiment Design," *Bulletin of the American Meteorological Society* 93, no. 4 (2012): 485–98. Richard H. Moss, Jae A. Edmonds, Kathy A. Hibbard, Martin R. Manning, Steven K. Rose, Detlef P. Van Vuuren, Timothy R. Carter, et al., "The Next Generation of Scenarios for Climate Change Research and Assessment," *Nature* 463, no. 7282 (2010): 747–56.

20. Rajendra K. Pachauri, Myles R. Allen, Vicente R. Barros, John Broome, Wolfgang Cramer, Renate Christ, John A. Church, et al., "Climate Change 2014: Synthesis Report," contribution of Working Groups I, II and III to the Fifth Assessment Report of the Intergovernmental Panel on Climate Change. IPCC, 2014.

21. National Research Council, *Climate Stabilization Targets: Emissions, Concentrations, and Impacts over Decades to Millennia* (Washington, D.C.: National Academies Press, 2011).

22. Christopher B. Field, Vicente R. Barros, Katharine J. Mach, Michael D. Mastrandrea, M. van Aalst, W. N. Adger, Douglas J. Arent, et al., "Technical Summary," in *Climate Change 2014: Impacts, Adaptation, and Vulnerability. Part A: Global and Sectoral Aspects. Contribution of Working Group II to the Fifth Assessment Report of the Intergovernmental Panel on Climate Change*, ed. Core Writing Team, R.K. Pachauri and L.A. Meyer (Cambridge: Cambridge University Press, 2014), 35–94.

23. Rebecca Kessler, "Stormwater Strategies: Cities Prepare Aging Infrastructure for Climate Change," *Environmental Health Perspectives* 119, no. 12 (2011): a514–19.

24. National Interagency Fire Center, "Welcome to the California Wildland Fire Coordinating Group (CWCG)," California Wildfire Coordinating Group. https://gacc.nifc.gov/oscc/cwcg/ (accessed June 1, 2017).

25. Feargus O'Sullivan, "Venice's Vast New Flood Barrier Is Almost Here," Citylab (2016). https://www.citylab.com/life/2016/09/venices-vast-new-flood-barrier-is-almost-here/498935/ (accessed June 7, 2017).

26. Gaurav Sarswat and Mohammad Arif Kamal, "A Critical Appraisal of Off-land Structures: A Futuristic Perspective," *Civil Engineering and Architecture* 2, no. 9 (2014): 323–29.

27. Paul J. Crutzen, "Albedo Enhancement by Stratospheric Sulfur Injections: A Contribution to Resolve a Policy Dilemma?" *Climatic Change* 77, no. 3 (2006): 211–20. David W. Keith, "Geoengineering the Climate: History and Prospect." *Annual Review of Energy and the Environment* 25, no. 1 (2000): 245–84. David G. Victor, "On the Regulation of Geoengineering," *Oxford Review of Economic Policy* 24, no. 2 (2008): 322–36.

28. UNFCCC, *Adoption of the Paris Agreement*. Report No. FCCC/CP/2015/L.9/Rev.1, http://unfccc.int/resource/docs/2015/cop21/eng/l09r01.pdf.

29. Joeri Rogelj, Michel Den Elzen, Niklas Höhne, Taryn Fransen, Hanna Fekete, Harald Winkler, Roberto Schaeffer, Fu Sha, Keywan Riahi, and Malte Meinshausen, "Paris Agreement Climate Proposals Need a Boost to Keep Warming Well Below 2 C," *Nature* 534, no. 7609 (2016): 631.

30. Glenn Althor, James E. M. Watson, and Richard A. Fuller, "Global Mismatch between Greenhouse Gas Emissions and the Burden of Climate Change," *Scientific Reports* 6 (2016): 20281.

CHAPTER 3

Deep Time and Landscape History
How Can Historical Particularity Be Translated?

ÖMÜR HARMANŞAH

The Challenges of the Anthropocene

Ecological disasters and global climate change have recently become matters of everyday concern around the world. From "breaking news" reports on hurricanes destroying cities to mass migrations caused by war, terrorism, and violence; from scientific reports on mass extinctions to apocalyptic feature films, we live in an era of ecological anxiety. In December 2017, a statement signed by 15,364 scientists from 184 countries published in the international journal *BioScience* issued a (second) warning to humanity that the current regimes have brought "substantial and irreversible harm" to the planet Earth.[1] As Bruno Latour recently pointed out, this is no longer an "ecological crisis" but an instability that is here to stay.[2] Crisis is a term that implies a temporary state of affairs that can be fixed by some technological intervention. Yet, environmental scientists are pointing out an irreversible process. For the first time in planetary history, the human species is considered a *geological agent* that introduced irreversible structural changes to the atmosphere, biosphere, hydrosphere, and lithosphere.[3] Yet far from bestowing on humanity the lion's share of power over the rest of the earth, as it is often assumed, the Anthropocene debate points to a sense of helplessness, to humanity's limitations in shaping its own deep history and future.

In this era of rapid change, violence, and ecological panic, a reconsideration of the distribution of agencies across the planet's history has become inevitable. Historians point to the implications of our moment of global anxiety on how we write history, while they underline the significance of being both a chronicler of the present and engaging responsibly with deep time. They equally urge us to reconsider historical fundamentals such as the nature/culture divide, prioritization of human over geological time, questions of freedom and agency,[4] and the idea of a shared future for humanity.[5] As academics, we are now reminded that a new sense of responsibility, a new ethics of collaboration, and—if I am allowed to be so optimistic—a new regime of care is demanded of us in doing our work as archaeologists, as historians, as anthropologists, as scholars in the humanities.[6] We must take this task very seriously.

Deep Time Leaks into the Present

Attending the workshop *Timescales: Ecological Temporalities across Disciplines* at the University of Pennsylvania in 2016, a fellow climate historian and I (a landscape archaeologist and architectural historian) were asked to respond to the challenge: how can historical particularity be translated in the context of the contemporary debates on the current ecological crisis, the new geological epoch of the Anthropocene and the challenges they pose on the writing of history? One of the revolutionary aspects of the debate around climate change and the Anthropocene is the weakening of the strict separation between historical time—that is, the temporality of historical writing—and the temporality of geological structures of the planet Earth, and the conviction that nonhuman, geological time is considered to be outside of history or relatively stable and immutable. *If the onset of the Anthropocene is a moment in which an unusual window is opened into the slow-moving processes of the mineral world, like an accidental and deep cut into the stratigraphy of the sediments of earth's history, and has demonstrated to us that the impact of the human species has always been at work as a geological agent in that history, what exactly would be the implication of this newfound understanding of deep time on historical accounts of the past?* If the Anthropocene can be defined as a narrow crack into deep time, I will suggest then that it offers a kind of temporality today in which the deep time leaks into the present.

If we define ourselves in the present in reference to a kind of historicity, and place ourselves within a linear sequence of recorded history and thus dwell in such historically informed identities, what would be the impact of the Anthropocene to this placement and such settled identities? If the strength of historical writing has always been its contextualization of historical events and processes, as well as their radical particularity and contingency, how does one account for an alternative ontology of historical time that goes beyond micro-histories of short-term political acts and engage with deep history? Similarly speaking, excavating archaeologists work with fine-grained material residues of the deep past and have developed meticulously refined forensic apparatuses to study past human lives through their engagement with resilient material things in increasingly precise spatial contexts.[7] How can all this archaeological and historical particularity be translated into the new sensitivities and new ontologies of time and the assemblage of more-than-human histories that have now become inescapable objects of study? We are then essentially talking about an expansion, an opening up of our conventional micropolitics of short-term political events and stratified material things.

The proposed epoch of the Anthropocene, it is suggested, has introduced an entirely new sense of temporality, allowing us to think in the long term with an increasingly widening sense of depth into the past and into the future. Studies of deep time are no longer restricted to the deep past, with the opening of debates around an expanded sense of the present and deep futures as well.[8] It can be pointed out that, in the debates around climate change and environmental crisis, the present has expanded its territory beyond the eventfulness and immediacy of the recent past and imminent future, beyond the myopic discourse of political history and beyond the practically minded vision of policy makers. These debates are opening the present to the impact of the Braudelian *longue durée,* the slow and forceful rhythm of geo-histories, one of the strongest historical paradigms of the Annales school of history.[9] Dipesh Chakrabarty expressed his reservation about the usefulness of the Braudelian *longue durée* in his influential article "The Climate of History" and states that the deep history of today's debates cannot maintain Braudel's vision of the long term, since it implies a cyclical version of history that is based on constant repetition.[10] In Braudel's *Mediterranean,* according to Chakrabarty, the environment still remains a silent and passive backdrop to the unfolding of human drama. We are in need of an understanding of landscapes and ecologies that are by no means static backdrops or dependable environments upon which cultural practice is inscribed, but are themselves agents that take part in the constitution of the world.[11]

Furthermore, the Anthropocene invites us to rethink the status of the human among other species, nonhuman beings, and ecologies, while it challenges the temporal sovereignty of "human time." The problem of deep time in the Anthropocene and the difficulty of translating historical particularity to an expanded view of the past, the present, and the future requires perhaps a new cosmology, new narratives of entangled histories (composed of human and nonhuman actors), and a new ontology of time that is not restricted to linear chronologies of human experience. I suggest that an archaeological way of thinking about landscapes and deep time, and notions of entangled materiality may be useful in reconciling the gap between the short-term vision of historical writing and the new ontologies of time in a changing climate. This genuine search within the Anthropocene debate for alternative narratives outside the Western paradigm increasingly brings together new materialist and posthuman discourses on the one hand ("the ontological turn")[12] and decolonial histories that undermine the slow violence of late capitalism and its disposable lives and landscapes on the other.[13] Deep histories of archaeological landscapes can pay attention to both the human and more-than-human histories on the planet's distinct regions,[14] while they effectively connect those entangled pasts to contemporary realities of ruination, contamination, and expulsion of human/animal/plant communities under the late capitalist governance of the planet. As much as I focus on questions of time and temporality, my argument is inevitably about landscapes and spatiality, as I have always been a spatial thinker.

Temporality of the Anthropocene

The proposal for the new geological epoch of the Anthropocene highlights the scientifically well-established understanding that the human species appears as a prominent agent of change in the earth's geology and ecosystems especially in the past two hundred years.[15] The current geological epoch, the Holocene, had started 11,700 years ago with the closing of the Last Glacial Period or the Pleistocene[16] and marks the beginning of the transition of human communities from a nomadic hunter-gatherer lifestyle to settled life supported by animal and plant domestication. There are heated debates about the beginning date for the Anthropocene: most emphasize the Industrial Revolution (usually designated as the invention of the steam engine in 1784 by James Watt),[17] particularly considering the global impact of the use of fossil fuels and the increased impact of greenhouse gasses. More recently proposed was the "Great Acceleration" after 1945 when the broader anthropogenic impact becomes unambiguously visible in the environmental record.[18] Therefore, *the debate around the Anthropocene is an intensely political and cultural debate*, involving the politics of sciences embroiled with scenarios of environmental crisis, the governance of the planet, and the role of capitalism and industrial modernity in the present predicament.[19]

As Dipesh Chakrabarty, Daniel Smail, and Catherine Malabou have stated, the question of the Anthropocene is also a question raised about narratives of conventional history and historical writing.[20] In these debates, the Anthropocene materializes both as a historical moment of dramatic change, as well as a recognizable geological era distinguished from what comes before (the Holocene) by material signatures in the stratigraphic structure of the earth's geology. What is at stake here is not simply a smooth transition or logical move from writing histories in a human timescale toward an understanding of the past in terms of deep time and writing environmentally conscious histories in the pace of slow geological time. Malabou points out that these two ways of thinking about the past have been epistemologically and ontologically incompatible and somehow will need to be reconciled within the project of the Anthropocene. How does one reconcile recorded history (largely dominated by short-term events and political actors) with the slow pace of geological change on the earth's surface (where the agency is distributed among human and nonhuman actors)? And to complicate things even further, what about all the other middle-range processes such as local vegetation histories, microclimatic episodes of fluctuating temperatures, rapid geomorphic changes in river basins and deforested landscapes accompanied by erosion, settlement histories on the regional scale, or technological changes in material culture?[21]

Take, for instance, ancient historian Eric Cline's recent book *1177 BC: The Year Civilization Collapsed*, a historical account of the socioeconomic and political world-system collapse at the end of the Late Bronze Age in the Eastern Mediterranean.[22] Archaeologists of the Bronze Age Mediterranean will be quick to tell you about this unique maritime-based trade network and connectivity that linked the Levantine Near East, Anatolia, the Aegean, and North Africa.[23] This seafaring trade network largely operated between coastal cities, and the primary agents of exchange were a plethora of entrepreneurial merchant ships traveling in a counterclockwise itinerary that was largely determined by the predominant currents in the Mediterranean and the ship technology at the time, which restricted seafaring primarily to coastal navigation.[24] This extraordinary mobility of entrepreneurial and transnational merchants was complemented, or perhaps shadowed, by the vibrant diplomatic exchanges and reciprocity between the great palaces of imperial powers, whose relationship was mediated by skillfully produced artworks and their "international style."[25] The systemic collapse of this interregional network in the early decades of the twelfth century BCE was clearly several centuries in the making, not under the control or agency of any political entity. The collapse of political institutions like the Hittite Empire, urban centers such as Ugarit and Emar, and the ceasing of the seafaring network cannot be explained through a myopic understanding of short-term political events such as the invasion of the so-called sea peoples, but is understood best when contextualized within largely systemic processes that involved ecological changes, climatic fluctuations, settlement shifts, political reorientations, effects of migration patterns between increasingly large urban centers and their hinterlands, the shift from bronze to iron technologies, changing lifestyles and cultural associations, and so on. Reading the Late Bronze Age collapse through the lens of the long term would have been appropriate in the manner of Peregrine Horden and Nicholas Purcell's *Corrupting Sea* or Cyprian Broodbank's *The Making of the Middle Sea*.[26] However Cline prefers the conventional historical angle and sensationalizes the collapse as an abrupt and cataclysmic event of political conspiracy and conflict, largely brought about by political actors and identifiable short-term events like "sea peoples" invasions, and characterized as an accident of history. The book in this sense is deliberately presented as a detective history with twists and turns, as the author admits.[27] This is actually a form of symbolic violence that writes out the agency of the subjects of history, in this case that of the seafaring merchants of the Mediterranean, the copper miners of the Cypriot mountains, the ivory craftsmen of the North Syrian workshops, the strong undercurrents of the Mediterranean Sea, and the diplomatic "greeting" gifts that traveled across the sea, who are the real actors of history. The Anthropocene is then a project that is critical of this violence and the centralization of political actors. In that sense, as has been suggested, it is already a postcolonial project.

My second example is the so-called Neolithic Revolution, which is proposed by some as the date for the onset of the Anthropocene. The paleoclimatologist William Ruddiman, for instance, associated the beginning of the Anthropocene with the emergence of farming and sedentary communities, i.e. the Neolithic Revolution, a process that started around 10,000 BCE in the Middle East following the short arid spell called the Younger Dryas.[28] The process of the spread of the Neolithic cultures of settled life and animal/plant domestication has always been characterized as the mastery of human communities over nature, as it involved literally its domestication, and therefore is identified as a dramatic change in our relationship to landscapes. This so-called agricultural revolution however is in fact a process that takes at least 3600 years of resilience and experimentation to be completed in the Middle East and was really not the result of any resourceful inventiveness of the human communities who had suddenly decided to settle and domesticate wheat and barley, sheep and goats. It was made possible by a combination of ecological and climatological factors, the end of the Ice Age, a certain change in the amount of carbon dioxide in the atmosphere, the acquisition of a certain stability of climate following the Younger Dryas and the challenge it offered to human communities, a certain degree of the warming of the planet, and the availability of certain species of wild plants and animals in particular landscapes. Most of those factors were outside the control of human communities: it is a 3600-year-long event without a decision maker, it is a creature without a head.

Johannes Fabian critiqued the construction of anthropological time through the denial of coevalness, "a persistent and systematic tendency to place the referents of anthropology in a time other than the present of the producer of anthropological discourse."[29] This applies even better to archaeological discourse in its treatment of its subjects. The primitive otherness of the archaeological subject then has to be produced as "spatialized time," a kind of temporality that is collapsed into a strict geo-temporal framework, positioned carefully in the deep past in a perfect box, so that the archaeological materials should not leak into the present. For instance, archaeologists often find themselves speaking about "the Neolithic" and there is an ambiguity of topographic imprecision and temporal bracketing that construes "the Neolithic" as a combined landscape of otherness, a time "much closer to nature."[30] What I suggest is that the Anthropocene offers a kind of temporality today in which deep time leaks into the present. Archaeological time allows us to think about an alternative, nonlinear temporality in which such a leakage becomes possible, through material remains and palimpsests in landscapes. One could call this "percolating time," that is, deep time leaking into the present through ongoing material entanglements and spatial proximities.[31] In similar terms, writing on Anthropocene landscapes, Elaine Gan, Anna Tsing, Heather Swanson, and Nils Bubandt speak of "traces of more-than-human histories through which ecologies are made and unmade" as *ghosts*, and suggest that "every landscape is haunted by past ways of life."[32]

Posthuman Landscapes

These two examples from ancient Mediterranean history bring us back to the debate about the necessity for rethinking the human condition in relation to other species of the world's ecosystems, nonhuman communities, things, and landscapes. This rethinking provokes a critique of the Enlightenment thought that placed humankind at the center of history and as the master of the natural environment. The Anthropocene urges us to think about such episodes of deep history in a way to incorporate alternative scales and rhythms of time and temporality, and allow alternative material relationships to come forward. In my own work, I have focused on landscape research and genealogies of place, as a way to bridge the local and the global, deep history and historical time. Landscapes are not dependable natural environments or passive backgrounds in front of which the drama of historical actors unfolds, such as this French historical painting from 1808 implies (Figure 3.1). What we see here is Hippolyte Lecomte's painting of Napoleon meeting with the ambas-

FIGURE 3.1. Landscape as pictorial setting/scenographia. Hippolyte Lecomte (1808), *Meeting of Napoleon with the Ambassadors of the Austrian Emperor near Leoben Steiermark on 7 April 1797.* Oil on canvas, 97 x 146 cm. Courtesy of Wikimedia Commons.

sadors of the Austrian Emperor near Leoben Steiermark on April 7, 1797, a very specific political event with political actors taking place in front of a *Welt-Landschaft*, a world landscape as a pictorial setting. A definitive, accurately identified historical event is enacted in a romantic landscape of this French history painting, where the specific place is somewhat reduced to a pictorial set design or *scenographia*. On the contrary, I consider landscapes as vibrant matter, and as agents of historical change. Compare this painting, for instance, with Gustave Courbet's "Source of the Loue" from 1864 (Figure 3.2) a geologically vibrant, eventful place that he depicted over and over again in a series of dense and dark paintings. The source of the river appears here as an evocative, powerful landscape that is about to swallow you whole.[33]

Dagomar Degroot's work with historical climatology gives us fascinating accounts of the interactions and intertwined nature of climate change and historical events.[34] When one moves away from conventional historical analysis to investigate the long-term trends against climate data, one is likely to discover a wealth of new relations in the past, previously unarticulated simply because those other scales

FIGURE 3.2. A vibrant, posthuman landscape. Gustave Courbet, *The Source of the Loue* (1864). Oil on canvas (107.3 x 137.5 cm). The Albright-Knox Art Gallery, Buffalo. Courtesy of Wikimedia Commons.

were not brought into the debate. However, in this model, are we not left, then, with two different, although linked, actors on the stage of history: the independently acting environment/climate of the planet Earth on the one side and humankind on the other? My sense is that the Anthropocene challenges the boundedness of these agents.

Gavin Lucas articulated how, through the use of temporal tropes such as chronology, absolute dates, and periodization, the time of the subjects of research is set apart from the time of the objects of research. This separation then allows the displaced time to be subject to free theoretical or interpretative play, unmoored from the concerns of heritage politics or stakeholders of history and place. This displacement can be described as an evacuation of a particular past to deep time in history. First, notable in this sense is Geoff Bailey's time perspectivism (the idea that different historical processes operate at radically different scales) and the materially anchored temporality in the form of archaeological palimpsests.[35] If we follow the advocates of symmetrical archaeology, a powerful archaeological paradigm in contemporary thought that makes critical use of relational ontologies and actor-network theory, we might highlight their concept of "percolating time," which deconstructs modernism's linear temporality and replaces it with an understanding that suggests time is a product of relationships, ongoing material entanglements, and spatial proximities.[36] Third, I would like to refer to Rebecca Schneider's work *Performance Remains*, where she argues for the theatrical nature of time, and introduces the concept of "syncopated time of re-enactment."[37] Syncopated time highlights the effective "againness of re-enactment," of performative recapturing of the past, in such a way that past events literally leak into the present (and the future) in the form of their doubling at moments of disturbance and the interruption of the present. The raw and fleshy reality of a moment from the deep past is brought into real time, as Schneider demonstrates to be the case in U.S. Civil War reenactments. Through this kind of leakage, remote temporal links are established, and the past, the present, and the future "inter(in)animate" each other in situations of cross-temporality.[38]

Fieldwork in the Anthropocene

The final point I would like to make is a call for fieldwork, but a new form of fieldwork: fieldwork as activism, fieldwork as political engagement, fieldwork as political ecology. There is a certain level of tension that has developed between the globally focused metropolitan theories of the Anthropocene and climate change, on the one hand, and the political-ecological struggles in various regions of the world, on the other. The politics of the Anthropocene and the concern for the environment today have brought about the variety of environmental injustices of the capitalist world order. The victims of these environmental injustices are not just animal and plant communities or specific ecologies like the oceans or the Arctic, but also human communities around the world that struggle to keep their rights to water, clean air, biodiversity, land, and place as a site of belonging and heritage. Archaeologists, anthropologists, and environmental scientists are often drawn into these political ecologies, as we might call them, to carry out urgent salvage operations. Political ecology as a field, then, offers a platform to engage with local situations of political conflict over the environment and its resources. What I am interested in is to understand the discursive reciprocal relationship of such localized political ecologies with the metropolitan theories of the environment that are emerging from academia today. How are the local conditions of ecology responding to, being affected by, and feeding back to those global perspectives? How precisely are the debates of climate change and the onset of a new geological epoch finding resonance, reflection, and response in regional imaginaries, in the rapidly changing contexts of human–land relationships, in situated experiences of time and temporality? I propose that field-work must be conceptualized as activism, as field practice, not simply to document political ecologies but to engage with and intervene in them. In order to do this, we need to be armed with a new set of methodologies to address place-based ecological conflicts. My sense is that our moment is one of activism, and it is one in which theory will emerge from the field.

In conclusion then, the challenges posed to historical disciplines by the current debates on ecological crisis and climate change are overwhelming and require us to work with alternative ontologies of time—a powerful call for a radical reconfiguration of the historical and the archaeological imagination. In a recent article entitled "Climate Change? Archaeology and Anthropocene," archaeologist Þóra Pétursdóttir wrote about "drift matter" of Anthropocene landscapes: marine debris that washes ashore, a mixed assemblage of natural and unnatural/anthropogenic materials, matter out of place. The meticulous archaeological study of the assemblage demonstrated "the deep and irreversible entanglement of 'natural' and 'cultural' materials, and the very vigorous processes—transformations, formations, associations—ongoing on the surface and in the depths of this chaotic compound."[39] "Drift matter" serves as a perfect metaphor for the story of things and landscapes in the Anthropocene. Likewise Anna Tsing has presented us the hopeful paradigm for Anthropocene landscapes in the *Mushroom at the End of the World*:

> Global landscapes today are strewn with this kind of ruin. Still, these places can be lively despite announcements of their death . . .

In a global state of precarity, we don't have choices other than looking for life in this ruin . . . Our first step is to bring back curiosity. Unencumbered by the simplifications of progress narratives, the knots and pulses of patchiness are there to explore.[40]

NOTES

1. William J. Ripple, Christopher Wolf, Thomas M. Newsome, Mauro Galetti, Mohammed Alamgir, Eileen Crist, Mahmoud I. Mahmoud, William F. Laurance, and 15,364 scientist signatories from 184 countries, "World Scientists' Warning to Humanity: A Second Notice," *BioScience* 67, no. 12 (December 1, 2017): 1026–28.

2. Bruno Latour, *Facing Gaia: Eight Lectures on the New Climatic Regime* (Cambridge: Polity Press, 2017), 7–8.

3. Erle C. Ellis, *Anthropocene: A Very Short Introduction* (Oxford: Oxford University Press, 2018).

4. Dipesh Chakrabarty, "The Climate of History: Four Theses," *Critical Inquiry* 35, no. 2 (2009): 197–222; Daniel Smail, *On Deep History and the Brain* (Berkeley: University of California Press, 2007).

5. Latour, *Facing Gaia*.

6. For solid examples of such cross-disciplinary collaborations, see Jason M. Kelly, Philip Scarpino, Helen Berry, James Syvitski, and Michel Meybeck, eds., *Rivers of the Anthropocene* (Oakland: University of California Press, 2018); Anna Tsing, Heather Anne Swanson, Elaine Gan, and Nils Bubandt eds. *Arts of Living on a Damaged Planet / Ghosts and Monsters of the Anthropocene* (Minneapolis: University of Minnesota Press, 2017).

7. Þóra Pétursdóttir, "Climate Change? Archaeology and Anthropocene," *Archaeological Dialogues* 24, no. 2 (2017): 175–205.

8. I refer here to the conference session organized by Karen Holmberg entitled "Archaeology of the Future (Time Keeps on Slipping . . .)" at the Theoretical Archaeology Group meetings (U.S.) at New York University, May 22–24, 2015.

9. Brent D. Shaw, "Challenging Braudel: A New Vision of the Mediterranean," *Journal of Roman Archaeology* 14 (2001): 419–53; J. H. Hexter, "Fernand Braudel and the *Monde Braudellien* . . . ," *Journal of Modern History* 44 (1972): 480–539.

10. Chakrabarty, "The Climate of History," 204–6.

11. Elaine Gan, Anna Tsing, Heather Swanson, Nils Bubandt, "Introduction: Haunted Landscapes of the Anthropocene," in Tsing et al., ed., *Arts of Living on a Damaged Planet* (Minneapolis: University of Minnesota Press, 2017), G1–14.

12. Jane Bennett, *Vibrant Matter: A Political Ecology of Things* (Durham, N.C.:

Duke University Press, 2010); Catherine Malabou, "Anthropocene, a New History?" Durham Castle Lecture Series delivered on 27 January 2016; Pétursdóttir, "Climate Change?"

13. Elizabeth A. Povinelli, *Geontologies: A Requiem to Late Liberalism* (Durham, N.C.: Duke University Press, 2016).

14. Tsing et al., *Arts of Living.*

15. Jeremy Davies, *The Birth of the Anthropocene* (Oakland: University of California Press, 2016). See also Clive Hamilton, Christophe Bonneuil, and François Gemenne, eds., *The Anthropocene and the Global Environmental Crisis: Rethinking Modernity in a New Epoch* (London: Routledge, 2015); Jedediah Purdy, *After Nature: A Politics for the Anthropocene* (Cambridge, Mass.: Harvard University Press, 2015). See discussion on the Anthropocene in the introduction to this volume.

16. Davies, *The Birth of the Anthropocene*, 3.

17. Davies, 43.

18. Clive Hamilton, Christophe Bonneuil, and François Gemenne, "Thinking the Anthropocene," in *The Anthropocene and the Global Environmental Crisis* (London: Routledge, 2015), 1.

19. Bruno Latour, *Politics of Nature: How to Bring the Sciences into Democracy*, trans. Catherine Porter (Cambridge: Harvard University Press, 2004); Latour, *Facing Gaia.*

20. Chakrabarty, "The Climate of History"; Smail, *On Deep History*; Malabou, "Anthropocene, a New History?"

21. Peregrine Horden and Nicholas Purcell, *The Corrupting Sea: A Study of the Mediterranean History* (Oxford: Blackwell, 2000).

22. Eric Cline, *1177 BC: The Year Civilization Collapsed* (Princeton, N.J.: Princeton University Press, 2014).

23. See, for example, Susan Sherratt, "The Mediterranean Economy: 'Globalization' at the End of Second Millennium BCE," in *Symbiosis, Symbolism, and the Power of the Past: Canaan, Ancient Israel, and Their Neighbors from the Late Bronze Age through Roman Palaestina*, eds. W. G. Dever and S. Gitin (Winona Lake, Ind.: Eisenbrauns, 2003), 37–62; Marian H. Feldman, *Diplomacy by Design: Luxury Arts and an "International Style" in the Ancient Near East, 1400–1200 BCE* (Chicago: University of Chicago Press, 2006); Ömür Harmanşah, "Review of Marian H. Feldman, *Diplomacy by Design: Luxury Arts and an "International Style" in the Ancient Near East, 1400–1200 BCE*," Art Bulletin 90, no. 1 (2008): 123–26; Ömür Harmanşah, *Cities and the Shaping of Memory in the Ancient Near East* (Cambridge: Cambridge University Press, 2013), 31–35; Cyprian Broodbank, *The Making of the Middle Sea: A History of the Mediterranean from the Beginning to the Emergence of the Classical World* (Oxford: Oxford University Press, 2013), 445–60.

24. Sherratt, "The Mediterranean Economy."

25. Feldman, *Diplomacy.*

26. Horden and Purcell, *Corrupting Sea*; Broodbank, *The Making of the Middle Sea*.

27. Cline, *1177 BC*, xvii.

28. William F. Ruddiman, "The Anthropogenic Greenhouse Era Began Thousands of Years Ago," *Climate Change* 61 (2003): 262–93. See also Ellis, *Anthropocene*, 75–102; Bruce D. Smith and Melinda A. Zeder, "The Onset of the Anthropocene," *Anthropocene* 4 (2013): 81–83.

29. Johannes Fabian, *Time and the Other: How Anthropology Makes Its Object* (New York: Columbia University Press, 1983), 34.

30. Gavin Lucas, *The Archaeology of Time* (London: Routledge, 2005), 125.

31. Christopher Witmore, "Vision, Media, Noise, and the Percolation of Time: Symmetrical Approaches to the Mediation of the Material World," *Journal of Material Culture* 11, no. 3 (2006): 267–92.

32. Gan et al., "Introduction," G1–2.

33. For longer discussion of Courbet's paintings, see Ömür Harmanşah, "Event, Place, Performance: Rock Reliefs and Spring Monuments in Anatolia," in *Of Rocks and Water: Towards an Archaeology of Place*, ed. Ömür Harmanşah, Joukowsky Institute Publications 5 (Oxford: Oxbow Books, 2014), 140–68.

34. Dagomar Degroot, "'Never Such Weather Known in These Seas': Climatic Fluctuations and the Anglo-Dutch Wars of the Seventeenth Century, 1652–1674," *Environment and History* 20, no. 2 (May 2014): 239–73; Dagomar Degroot, "Exploring the North in a Changing Climate: The Little Ice Age and the Journals of Henry Hudson, 1607–1611," *Journal of Northern Studies* 9, no. 1 (2015): 69–91; Dagomar Degroot, "Testing the Limits of Climate History: The Quest for a Northeast Passage during the Little Ice Age, 1594–1597," *Journal of Interdisciplinary History* 45, no. 4 (Spring 2015): 459–84.

35. Geoff N. Bailey, "Time Perspectives, Palimpsests, and the Archaeology of Time," *Journal of Anthropological Archaeology* 26 (2007): 198–223.

36. Witmore, "Vision, Media, Noise, and the Percolation of Time."

37. Rebecca Schneider, *Performing Remains: Art and War in Times of Theatrical Reenactment* (New York: Routledge, 2011).

38. Schneider, 135.

39. Pétursdóttir, "Climate Change?" 177.

40. Anna Lowenhaupt Tsing, *The Mushroom at the End of the World: On the Possibility of Life in Capitalist Ruins* (Princeton, N.J.: Princeton University Press, 2015), 6.

ETUDE 1
A Period of Animate Existence

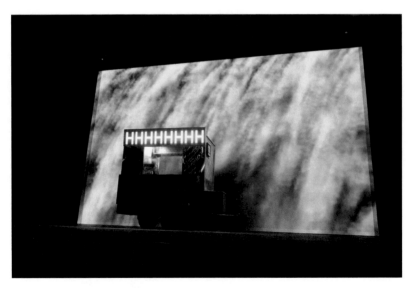

PLATE E1.1. A halal cart speaks in its period of animate existence in front of pouring rain. From *A Period of Animate Existence* (2017) on September 22, 2017. Photograph by J. J. Tiziou; courtesy of the artist.

PLATE E1.2. A grandmother and child consider their periods of animate existence. From *A Period of Animate Existence* (2017) on September 22, 2017. Photograph by J. J. Tiziou; courtesy of the artist.

PLATE E1.3. A halal cart invites collective song in front of a carbon-saturated sunset. From *A Period of Animate Existence* (2017) on September 22, 2017. Photograph by J. J. Tiziou; courtesy of the artist.

Staging Climate

A Period of Animate Existence and the Global Imaginary

MARCIA FERGUSON

Pig Iron's ambitious work in progress *A Period of Animate Existence* contributed rich artistic and performative provocation to the *Timescales* conference, held at the University of Pennsylvania in October 2016 under the auspices of the Penn Program in the Environmental Humanities (PPEH). I was invited to moderate a postperformance discussion of the artistic, personal, and political interventions suggested by Pig Iron's intimate and experimental staging of a small section of the larger performance piece, which had its premiere at the Philadelphia Fringe Festival in September 2017. As presentations and discussions at the conference clearly indicated, the field of the environmental humanities is in the process of determining its own terrain and boundaries, even as it wrestles with the compelling depth and breadth of its subject matter. As Ursula K. Heise writes, "The environmental humanities . . . envision ecological crises fundamentally as questions of socioeconomic inequality, cultural difference, and divergent histories, values, and ethical frameworks."[1] This common ground, to vividly imagine (or "envision") the effects of environmental crises on human beings, fundamentally unites the environmental humanities with *A Period of Animate Existence*.

Director Dan Rothenberg, composer Troy Herion, and set designer Mimi Lien were collectively the 2016–17 artists-in-residence at PPEH. Their groundbreaking piece connects academic inquiry with formal artistic practice, its incubation taking place at a nexus of restless disciplinary juxtaposition and theatrical combination prompted by global climate change. *A Period of Animate Existence* (henceforward, *PAE*) and environmental humanities, both featuring materials and approaches that engage distinct theories and practices drawn from science, art, theater, and environment, reside on a spectrum of productive categorical disruption. By privileging the elastic theme of timescales as a formal structure, the conference provided a productive focus for the proceedings, which included this performance, along with other diverse presentations, related to time and climate change. Pig Iron's artistic residency at the PPEH, situated as it is within a science-forward academic field, reflects the multidisciplinarity inherent in both the artistic project and the scholarly pursuit; but such novel combinations of theory and practice, humanities and science, logic and art, often trouble traditional boundaries by disrupting long-held, formative distinctions between disciplines and practices.

The work-in-progress excerpt in fact prompted doubt in the mind of at least one audience member, who immediately questioned the suitability of its inclusion at the conference. The postperformance discussion began with the query, "What does this have to do with environmental humanities?" This provocative question, for me, was a mark of *PAE*'s uncanny power. By immediately instigating a productive conversation about the progressive relationship of the arts and sciences in their respective engagements with the environment, the performance quickly conjured the field's most central considerations: What do you think about the future? Who will live to see it? How can we speak about it? Will the conversation between humanity and nature, which we have expressed through the arts and sciences for millennia, continue or end? If it ends, will all the advances of humanity, in every epoch, mean anything? Is the universe alive, and will it silence us through the extinction of our species? In which case, what is the meaning of time as we experience it in the present—specifically, this time, our shared time on the planet? The questions spin into reflections on artistic practice itself, as an "appropriate" experimental site for critical inquiry. They also point the way toward the multifaceted tonal effect of *PAE,* in which its relationship to the concerns of the conference (and indeed to the field of environmental humanities) becomes clear.

To return to the conference participant's question: What does theater have to with environmental humanities? In its seasonal rhythms, in every dimension of its materiality, theater has always addressed and engaged the human drama of our location on the planet. From the creation dramas of ancient Egypt to the natural cataclysms of eighteenth-century *Sturm und Drang* plays, from the elemental Furies of Aeschylus (who reside at the center of the earth, drawn forth to its surface by matricidal blood) to Jonson's and Moliere's comedies, which turn on the inexorable power of nature's humors (phlegm, blood, black bile, and yellow bile) to determine human personality and destiny; in theatrical practices and traditions such as Kathakali dance theater, Balinese shadow puppetry, and Japanese Noh drama, the innate functions of nature and the movements of the planets are dramaturgically evident in the structure, themes, locations, and plots of plays, across cultures and epochs. Whole theories and methods of contemporary practice and pedagogy, and generations of playwrights, have been inspired by an ongoing engagement with specifically environmental concerns.

Conversely, what does environmental humanities have to do with theater? Environment, space, and place are core values that shape and connect theatrical art, scholarship, and history. The formative Western Aristotelian definition of theater as a representation of an action is at once prescriptive and unbounded. Greek and Elizabethan drama did not limit the scope of representations to the observable but became dramas of spatial verticality, in which an imagined cosmos on high, inhabited by transcendent deities, could be invoked to come to the aid of sublunar, environmentally dependent humanity. The history of theater architecture is also a history of environmental pressures shaping the spectatorial experience. These same pressures became the focus of much theater and theory of the twentieth and twenty-first centuries. For example, environmental considerations are key to Brecht's experiments in alienation, which use space in the form of distanciation to inculcate social and political awareness in audiences (the *verfremdungseffekt*), playwriting, acting, and stagecraft. In the 1960s, Richard Schechner and Jerzy Grotowski sought to minimize the spatial relationships between audiences and actors, Bread and Puppet Theater reclaimed public space for processional theater (a reiteration of the space and movement characteristic of medieval cycle plays), and more recently, Anne Bogart and Tina Landau developed "Viewpoints," an influential, spatially oriented approach to actor training and performance. Environmental theater of all kinds continues to privilege and leverage space and environment as primary tools in the creation of diverse theatrical, social, and political effects too numerous to encompass here.

PAE's aim to represent (in the words of director Dan Rothenberg) "the very big within the very small, and vice versa," links it to this long lineage of theatrical explorations of space.[2] According to Rothenberg, Gustav Mahler's remark that a symphony "must be like the world, containing everything,"[3] became the formal touchstone of their creative process. Recalling his search for a suitable form, one that gestures toward its own inability to contain the vastness at the heart of its subject, composer Troy Herion cites as his inspiration the nineteenth-century five-movement symphonic structure. As Rothenberg describes it, the intersection of theme and form that characterizes this piece is "a deliberate effort to get at something that's too large to get your head around, by coming at it from five very different angles."[4] *PAE* indeed gets at the concerns with time and climate motivating the environmental humanities by leveraging theater's paradoxical capacity to represent time and space as at once imagined and real, boundless and limited, yet always confined to the real-time presence of performers and audiences.

The final version presented at the 2017 Philadelphia Fringe Festival was composed of five movements. The first conveyed the sense of incomprehensible largeness ("too large to get your head around") through design and music: monumental chords, projections of the five extinctions, and the slow, dramatic emergence of a spectacular, stage-filling object resembling a black planet turned spinning cone. The second movement pulled into tighter focus with a song, "Humans," danced and sung by a chorus of individuals who converged into unity through ensemble choreography. The third movement featured the musings of a Halal food truck, which advertised its offerings both edible and philosophical across its digital message board. Movement four was a pageant of elders and youngsters, with light dialogue featuring a conversation between a grandmother and granddaughter that was at once humorous and cosmic. The piece ended with a movement containing a final offering of projections about trilobites, the most enduring animals on the planet (they lasted for 270 million years before extinction). Brightly uniformed wrestlers, and trilobite-costumed singers, concluded the piece with Herion's rousing and infectious score.[5]

Maurya Wickstrom, in her useful recent overview of contemporary theater practice and scholarship centering on theatrical time, states

> although theatre as a medium is strikingly fluent in and fluid with temporality, we have not perhaps been as engaged with temporality in its own right as some other disciplines have been . . . *most* of this work has not been engaged with the question of temporality per se, with opening out the very meaning and practices of temporality itself. Increasingly, my interest, and the interest of a growing number of scholars and theatre artists, is in how theatre and performance are engaging with time in ways that do just this, guided by explorations undertaken through a variety of philosophical and theoretical apertures which influence political thinking in unfamiliar ways.[6]

PAE, I would argue, is at the forefront of such experimentations with theatrical time and space, in taking up time itself, and the meaning of individual lifespans both cosmic and human within it, as the site of its multivalent theatrical exploration. In the smaller version of *PAE* presented at the conference, the pared-down, work-in-progress performance used music, poetic dialogue, choreography, and a multigenerational cast to aesthetically and economically engage the key questions of the Timescales conference: that is, what it means to be human in an environment that is peacefully (one might even say, rhythmically) familiar, yet also evolving in disturbing, beautiful, and unfamiliar ways—perhaps toward its own ending. The excerpt featured an elegiac tribe of elders invoking the rhythms and rituals of daily life through repetitive movement, gesture, and speech. The weather, mealtimes, sleeping and waking, among other things, were the subjects of this group performance. Neither serious meditation nor cynical parody, they nonetheless achieved a tonality that resides somewhere in between those two poles, the way a recognizable, average conversation with an elder can also gesture toward the infinite. The quotidian words and actions are at once timeless, beautiful, and relatable because, one way or another, they pulse beneath every human life, regardless of location, epoch, or culture. The vision of so many older individuals onstage at once is also oddly thrilling, on the one hand for its novelty (rarely in our youth-centric culture do we see so many elders in performance at once), and on the other because of the symbolic resonance of their individual lives, which circle out to engage with the thematic materials spanning the piece across timescales.

Importantly, this symphonic performance movement also featured a tribe of children, moving and speaking in parallel, yet more disruptive, chaotic, and therefore equally familiar counterpoint (that is, the innate rhythms of a healthy childhood are, like the slower rhythms of the elders, inherently recognizable in human experience). They darted in and out among the more structured movements of their elders even as they looked to them (for guidance?), running, skipping, weaving and dancing among them, suggesting the infinite ways the young model their survival on the skills and protection of the elders, yet also challenge, evolve, and innovate upon received tradition as a matter of course. The effect of this surprisingly energetic (and intimate) playing, comprising a casual yet symbolic dance of life, was indescribably moving. Without traditional character development or plot structure, the spectacle of the literal interweaving of old and young was at once philosophical, moody, and playful, and moved the project another step toward its emotional and conceptual peak. This short symbolic movement gave spectators a surprising sense of entirety, as if we had witnessed whole life cycles, seasons, and generations unfolding and interweaving, with the greatest possible poetry and economy.

The conference excerpt also tidily demonstrated theater's unique position as the ideal, living location for experimentations with time, always telescoped into the present moment. Indeed, grasping theater's fungible internal timeframe is key to understanding the disparate movements of the larger performance of *A Period of Animate Existence* as a collective whole. Citing Toshiki Okada, a contemporary playwright whose work resides on a spectrum between a deceptive naturalism and a conversational hyperrealism, director Dan Rothenberg comments that Okada "writes very political plays but they take place in the minds of ordinary people, with very ordinary concerns. So sometimes we are following that lead, trying to get at the dull feelings of worry that pulse beneath the everyday. And other times we want to evoke the feelings of awe that arise when contemplating enormous scales of time and space."[7] Wickstrom also isolates this theatrical quality in her discussion of Matthew Wagner's 2012 book *Shakespeare, Theatre, and Time*,[8] which, she claims, "insists throughout on the unruliness of time in the theater, its refusal to obey the clock."[9]

Two other productions serve to extend and contextualize the theatrical and philosophical impact of shifts of time that characterize *PAE*. The anxiety of the everyday characterizes the time-warped dramatic action of Cecilia Corrigan's experimental political play *Motherland*,[10] also produced in 2017, as part of the This Is Not Normal festival at New York's Brick Theatre. The play centers on a talented young intern for Hilary Clinton's presidential campaign, who is thrust into a twilight world on the set of the 1990s sitcom *Seinfeld*, into which Clinton has retreated from the complications of an overwhelming present. Like a spy from another era, by adopting the everyday habits and rhythms of a 1990s television show (such as landline phones and relentless one-liners), the intern seeks to extricate her candidate from a previous "simpler" time to a demanding, complex present, with mixed results.

In Anne Washburne's extraordinary *Mr. Burns, A Post-Electric Play*, the setting for the three acts is described in terms of time only: "The very near future, Then 7 years after that, Then 75 years after *that*."[11] In this play, the few remaining inhabitants of a postapocalyptic landscape rely on retellings of plots from episodes of the animated sitcom *The Simpsons* to salvage a sense of communal human culture. As the onstage time progresses, these retellings and performances take on a blurred quality, their morphing demonstrating the traces of a lapsed collective memory, which results in "new" versions of original characters and plots. *The Simpsons*, here, becomes a classical precedent, recalled through a performance of remembrance, aping the way, for example, Renaissance and Elizabethan dramas reclaimed ancient Greek and Roman texts for their audiences.

Like *PAE*, these works use the familiar rhythms of the everyday to evoke vast differences in space and time. While the separate movements of Pig Iron's piece are freestanding, each also resonates with the other, so that the individuality of a movement, and of a life, is understood in terms of a collective, cosmic whole. Like children running through slow-moving elders, this inclusive collectivity can make for chaotic stage pictures that (to use Wagner's term) insist on interruption and unruliness as the natural order of things. Paradoxically, dramatic timescales here enthrall precisely through the quotidian rhythms of that familiar unruliness. *A Period of Animate Existence* theatrically articulates the collisions and collusions between the same infinite and cosmic, finite and worldly, quotidian and eternal spaces in human experience that environmental humanities intellectually engages. Each harnesses and enlivens the creative vitality at the heart of an ongoing human moral engagement with the forces of nature and time, especially critical today.

NOTES

1. Ursula K. Heise, "Introduction," in *The Routledge Companion to the Environmental Humanities*, ed. Ursula K. Heise, Jon Christensen, and Michelle Nieman, 1–10 (New York: Routledge, 2017).

2. "Textbook Definition of Life: Interview with Dan Rothenberg of Pig Iron," *Fringearts* blog, July 13, 2017, http://fringearts.com/tag/a-period-of-animate -existence.

3. "Textbook Definition of Life."

4. "Textbook Definition of Life."

5. For a full description of the final performance, see Toby Zinman's review, "Hello, earthlings. Also, goodbye," *Broad Street Review*, September 23, 2017, https://www.broadstreetreview.com/theater/philly-fringe-2016-pig-iron-theatre -companys-a-period-of-animate-existence#.

6. Maurya Wickstrom, "Thinking about Temporality and Theatre," *The Journal of American Drama and Theatre* 28, no. 1 (Winter 2016): 1; http://jadtjournal .org/2016/03/23/thinking-about-temporality-and-theatre/.

7. "Textbook Definition of Life."

8. Matthew Wagner, *Shakespeare, Theatre, and Time* (New York: Routledge, 2012).

9. Wickstrom, 1–2.

10. Cecilia Corrigan, *Motherland*, performance at Brick Theatre, New York, June 27, 2017.

11. Anne Washburne, *Mr. Burns, A Post-Electric Play* (London: Oberon Books, 2014).

CHAPTER 5

A Period of Animate Existence

As artists in residence at the Penn Program in Environmental Humanities, Pig Iron Theatre Company has developed a new work of symphonic theater—a meditation on life cycles and planetary cycles in a time of rapid ecological and technological change.

A Period of Animate Existence is a synthesis of theater, music, and design, conceived and coauthored by composer Troy Herion, designer Mimi Lien, and director Dan Rothenberg.

We are writing for children, elders, classical musicians, choirs, machines, physical actors, and dynamic environments.

PROJECT DESCRIPTION

Echoing the character and pacing of a nineteenth-century symphony, the story will weave together perspectives of children, elders, and machines in five staged movements.

Like a symphony, the movements are unrelated narratively, yet they complement each other sequentially in mood, tempo, and texture to produce a cumulative effect.

APPROXIMATE RUNNING TIME: 110 MINUTES WITHOUT INTERMISSION

MOVEMENT 1
Con moto maestoso (with majestic motion)
An orchestra performs an evocative prelude as a ballet of primordial scenic forms moves through light and projections. This is a "plastic theater" reminiscent of the work of Josef Svoboda or the interludes in Terrence Malick's *Tree of Life.* The audience is disoriented and wonders if what they are watching is celestial, mechanical, or biological.

MOVEMENT 2
Andante cantabile (flowing and songlike)
"Humans have no advantage over animals." The stage overflows with dozens of people. Together in groups large and small they sing about planetary cycles, reproduction, death, and inheritance.

MOVEMENT 3
Scherzando (joking)
A single halal cart with scrolling LED screen gradually gains consciousness and considers its role in the universe.

MOVEMENT 4
Recitative (rhythms of ordinary speech)
A young girl interviews her mother and dying grandmother about the future. Their words take on musical qualities and become a children's pageant at the end of time. Songs and carnival floats alternate with a child's investigation into the nature of time.

MOVEMENT 5
Danza di rocce (dance of rocks)
A visualization of Henri Bergson's "élan vital," the force that separates the living from the nonliving. A choir sings in hushed tones behind tableaux of muscular bodies struggling.

INTERLUDES
Between movements are short, minimalist encounters between children, elders, and musicians.

WE FIND OURSELVES IN A PERILOUS TIME

One that is called the Sixth Extinction, an era in which we foresee the loss of 20 to 50 percent of all living species on earth. The gravity of these issues has entered mainstream consciousness, affecting our politics, media, and ultimately our individual beliefs about the trajectory of life.

How do we contemplate the future in such a movement?

A PERIOD OF ANIMATE EXISTENCE IS OUR INQUIRY INTO THIS GREAT DISRUPTION

What is this effort for life to go on living: the language used by one generation to address another; the difference between the animate and the inanimate; and the visceral feeling that the force of life itself— what Henri Bergson called "the élan vital"— permeates our notions of minerals, plants, animals, people, and time.

NOTES ON FORM

We looked at art works that encouraged contemplation and ecstasy—works and experiences that suggested the dissolution of self and the illusion of timelessness. In performance or time-based art, a lot of this work is either "durational" or "minimalist." We decided to go another way: to use short, pungent theatrical events to achieve feelings of suspension and transcendence.

ON ARCHITECTURE AND FORM
MIMI LIEN
SET DESIGNER

Throughout architectural history, the design of cathedrals has aspired to inspire the presence of the divine through volumes of space, and the orchestration of light. Now, in the time when cathedrals are no longer being built (at least not the way they used to be), one could still say that an ideal project for a particular kind of architect might be a structure which eschews practical function, but rather engenders a certain kind of contemplation, a questioning of our perceived place in the universe. In a way, this is the purest expression of architectural form.

ON SYMPHONIES
TROY HERION
COMPOSER

In the nineteenth century, the symphony was the largest and most ambitious form of instrumental music used by classical composers. Traditionally, the symphony breaks into four separate movements, each one highly individual with its own texture, mood, and tempo. Melodies from one movement rarely show up in another movement. Instead there is a strong sense of contrast from one movement to the next, juxtaposing loud and soft, fast and slow, aggressive and emotional, serious and comical. Overall it is a form of accumulation, where highly contrasting points of views are presented together as a whole. The symphony has a reputation for grand, sweeping statements. As Gustav Mahler famously declared, "A symphony must be like the world. It must contain everything."

ON THEATER AND THE SACRED
DAN ROTHENBERG
DIRECTOR

My wife was raised as a person of faith. When we were first dating we drove across Ireland and she pulled over the car, in tears, and declared "you have no sense of the divine in man." "Oh come on!" I insisted. "You know I do; I just use different words for it."

It seems to me that theater is an unlikely vehicle for communication about "the sacred." A volume of light or a pattern of music approach the "sacred" more easily than the spectacle of human bodies struggling, worrying, enacting our small dramas. I spent a long time identifying as secular and skeptical, but recently I've been forced to confront my capacity for prayer. To try to deploy people—us fleshy, imperfect, distracting humans—on a stage and still evoke "the sacred": or at least fundamental forces in the universe has been part of the challenge here.

COLLABORATORS' CONNECTION
TO THE WORK

We are collaborating with volunteer choristers ages 7 to 81. It is rare to see these bodies on stage, and even more rare to see them together.

YOSHIFUMI NOMURA
CHORISTER, AGE 23

I like when we sing "sleep it off. Begin again. It's been like this. For a long time." While in college, I came to the realization that life is pointless and is just an illusion. A lot of my friends in college are more optimistic, and even my family members make fun of me how cynical and nihilistic I am. So when I first read these two lines, I felt so embraced.

PAT JORDAN
CHORISTER, AGE 70

"Many years—four billion years—is not enough." Reflection, fine as it can be, is looking back on some other "very moment." And concern with the future, while it's necessary, can become compulsive—an obsession that excludes appreciation of the immediate. REMINDER TO PAT: "This very moment" is IT.

UMER PIRACHA
CHORISTER, AGE 32

We sing, "the human soul—so majestic." It's precisely the subjective, conscious experience that stands becoming irrelevant because of pattern recognition technologies and AI. I think we'd do well to treat the experience of the soul on its own terms even if it technically doesn't exist, using mythology that helps preserve it, given what's coming.

ZANNA YOSHIDA
CHORISTER, AGE 23

In a time where the world seems to be crumbling around us, dwelling on the darkness won't help me or anyone else. We need to continue to create, love, celebrate, feel, and move. Each one of us will keep breathing and living until we don't. Why worry about that moment before it comes around?

COLLABORATORS' CONNECTION TO THE WORK

NAIA GOMES
CHORISTER, AGE 9

When I'm eighty, I want to be like my grandpa. He still works. He likes his job. So I'm hoping I can be like him. At school we learned what scientists used to measure weather like hurricanes and tornados, and in Texas there's a place called Tornado Alley. There are all these ditches, so when a tornado hits, people can just jump into a ditch. My teacher told us about that.

ARLENE HILTON
CHORISTER, AGE 81

When I think of the future, I think of my consciousness moving from the earthly awareness and awakening into another space or dimension, where time is simultaneous and my consciousness has the ability to perceive with spectacular fluidity and within multiple dimensions. I will merge with my higher self, and I will be "home." I feel it will be an adventure; looking forward to it.

NANCY BOYKIN
ACTOR, AGE 67

I play the last Grandma of All Time. Although I have not yet passed from this animate world, I can feel that I am getting closer to an unknown other world. Grandma has moved past optimism and pessimism. She is part of the past, the present, and the future, so there is nothing to fear. I am an optimist. Although I have great concerns about the immediate future, I trust in the hope and goodness of the next generation.

MEL KRODMAN
ACTOR/CREATOR, AGE 37

I don't view the end of humanity as a doomsday, rather just as the way the whole thing goes. I don't know if that's a traditional view of optimism! My character in this piece is an alternate-universe version of me, in which I got married at twenty-seven and had a daughter at thirty. I think she's a pragmatist. She is willing to leave room for optimism, and I don't think she assumes the worst—I just think Melissa feels the need to prepare for the fact that the worst might happen.

COLLABORATORS' CONNECTION
TO THE WORK

JENN KIDWELL
ACTOR/CREATOR, AGE 38

I don't know how we could understand our universe to be anything other than "alive," because that's how we contextualize motion, heat, movement. Imagining us humans as the atoms of the universe and thinking about how chaotic our movements seem from a distance is a fun exercise. When I think of the future, I wonder if we're ever going to be able to reckon with our own insignificance and exist in harmony with the rest of the universe, as opposed to living to master it.

MARGALIT EISENSTEIN
ACTOR, AGE 11

Climate change is changing the weather and the temperature of the earth. I try not to worry about that now, but stuff keeps coming. My friends and I do talk about the future, like what kind of person we're going to be, but we don't really talk about the problems that might happen because there's only a small chance they might happen. Or a big chance, but we don't know.

DONALD NALLY
CONDUCTOR, AGE 56

My future is finite, and, at fifty-six, I can see that end. Then there is the future of humans and I can also see that end; we are committing to the finality through a corporate agreement. I find a certain comfort in knowing that the universe doesn't think—it just is—and that it doesn't care about the end of my life or the extinction of humans, because it can't. It just expands, as time and space recede into the past, which is also, in a way, an illusion of observation.

LOREN SHAW
COSTUME DESIGNER, AGE 35

Is the Universe alive? Yes, duh. Yes . . . it's definitely not not-alive. It's not dead. There is a grand consciousness and all these things are happening all the time that we aren't even aware of. I guess maybe we could just be decomposing? Did the universe die at some point and we're just experiencing the aftermath? I hated my science teacher in high school, but I should have paid more attention.

2
school bell
finger roll/superball mallet
improvise tremolo scratches in rests
LOOP 2
fling outward
breathy
deviate timing and pitch, interact with flute
mute in
without mouthpiece, pitchless
wah
half stop
tongue stop
tongue stop
muted
pressure!
palm mute
tasto
spicc.
spicc.
half pressure l.h., free pitch
tune E as low as possible,
pluck + rattle against A string
Fl.
multiphonic (F/C), blend with saxophone
Ob.
Cl.
multiphonic (Bb), blend with oboe
Ten. Sax.
remove tuning slide, cover 3/4 of way
Hn.
(in Bb) play E open w/ 3rd valve 1/2 down
Tpt.
multiphonic
Tbn.
harmon mute -> open
Tbn.
multiphonic
harmon mute -> open
Remove bell
brush on hihat/stand leg
Perc.
upside-down snare w/ paper
Perc.
high pressure, ascending pitch
Vln. I
palm mute finger pick, accel.
Vln. II
high harmonics, same figure as trumpet
Vla.
sliding harmonics / high notes
Vc.
sliding harmonics
Db.
LOOP 1
breathy
Flute
pitch up, alternate fingerings
Oboe
slaptongue/
multiphonics
Clarinet in Bb
multiphonic Bb
Tenor Saxophone
wah
half stop
Horn in F
(in C) tongue stop
Trumpet in Bb
Trombone
Trombone

CHAPTER 6

Conversations with Dan Rothenberg, Director of *A Period of Animate Existence*

BETHANY WIGGIN

While an Artist in Residence at the Penn Program in Environmental Humanities, Dan Rothenberg, who directed *A Period of Animate Existence*, and I, who served as one of two dramaturgs, talked together, a lot. We walked and talked, we talked over food and drink, we talked on the phone, we talked on camera. (One time we even recorded our conversation, but we were especially tired that day, and the conversation was pretty boring.) Of course we also talk a lot over email.

The conversation included here occurred via e-mails after the September 2017 world premiere of *PAE*, over the course of a few weeks in early 2018. We conducted it especially with the *Timescales* book in mind, desirous of the space and time to reflect on the many chitchats that had given the heartbeat to our collaboration on *PAE*.

Rothenberg is a founding artistic director of the Pig Iron Theatre Company, and I wanted to work with the Company and on *PAE* not least for the ways in which the devised performance methods pioneered by Pig Iron intersect with the spirit of the arts + sciences brews on the burners at www.ppehlab.org. We're always trying to cook up experiments that refuse to reveal if they're art or science. My chitchats with Dan, like others in this book, point toward a future and ways of apprehending a changing world that are by necessity and by design improvisational.

BETHANY WIGGIN: You've been making collaborative experimental theater for a while now (two decades, right?). Why was "now" the moment to create this ambitious musical hybrid about ecology and climate change?

DAN ROTHENBERG: I think [composer] Troy [Herion] puts it best when he says that "these dire predictions" have "become part of mainstream consciousness." At first the work was going to be a kind of a sharp joke, a scolding Tom Tomorrow cartoon writ large: kids haranguing their elders for despoiling the planet. I think this was a kind of cartoon response to the strange spectacle of the Republican party continuing to deny climate change science well into the twenty-first century—my understanding is that they are the only major political party *in the world* to take this position. Meanwhile, this party takes on the mantle of "protector of the family"—I kept wondering: what are you going to tell your grandchildren? Do you really think they are going to buy "we didn't know"?

But I put off making that piece for ten years, probably because it was so narrow, so snarky, such propaganda—more suited to a four-panel cartoon than a two-hour performance.

Troy expanded the horizons for the piece; he expanded them as far as they go, to the trajectory of life in the universe. To this spiritual question, "what is the difference between alive and not-alive?"—a question that is endlessly rich, both for scientists and philosophers.

I think some people were offended that the resulting piece was missing a "call to action." But that was part of the point. The information is out there. And calls to action are out there. Another call to action, in the same tone, doesn't seem necessary to me.

BW: Can you talk about what it was like to collaborate with nine- and ten-year-olds to make this piece? It has some dark scenes (when Mel, who plays a mother, completely melts down after the Grandma's death, for example). How did the kids handle that? How did you?

DR: In fact we had kids as young as seven in the piece. I think you're asking about the ethical implications of sharing this information with very young people. I have a two-year-old, and I know, very intimately, that she does not yet understand what death is.

I blundered into this ethical dilemma during our very first exploratory workshop. We invited a children's chorus of thirty kids to come work with us for a day (and we learned quickly that thirty children is way, way too many). After a couple of hours of singing, screaming, and rolling on the floor, it was time to wrap up. One little girl raised her hand. "Yes?" I asked. "I have a question. So what *are* we going to do, you know, about climate change?" My blood ran a little cold. "I don't know," I said. "I think that all of us adults, we are hoping that people like *you*, people from your generation, will be the ones to figure it out." The children all went silent. "Thanks so much for coming and playing with us today!" I concluded.

I don't think I caused too much panic among the youth that day—I imagine those kids have other authority figures they trust a lot more than me. But I realized I had to think this through, and talk to the parents about it, before I started up with the project again. I knew that I didn't want or need the child performers to do any dying on stage (though they would have been thrilled to do that, I think). So I explained to the parents, and then to the children, that this play was going to have some very sad parts in it, and dark parts in it. That their part of the play imagined a little girl losing her grandmother, and that the play itself contemplates the possibility that someday all living things will be dead. They took it in stride. I suppose that kids are more in touch with their dog brains than we are: they are probably more frightened when they see the adults frightened, and we didn't seem frightened.

Now that I've said that, I'm wondering if that's the whole problem, from a policy perspective: we don't feel this crisis in our bones, because we look around, and we don't see anything like panic.

BW: I'm thinking about what you said about kids being more in touch with their "dog brains," and when they don't see adults looking or acting scared, they aren't either. While making *PAE* you also talked a lot with environmental scholars across disciplines. What unites a lot of them (or maybe us) are feelings of despondency that can border on feeling scared. What was it like to work with scholars?

DR: It seems to me that both artists and intellectuals are more siloed here in the States than in other countries I've visited. I don't have any data on this, of course—it's just a collection of impressions. In Latin America, it feels like artists have a stake in some of the revolutionary language and thinking that established so many of the nation-states in that part of the world. It seems like artists have a role to play in the political life of the nation. In Europe, of course there is another scale of state support for artistic endeavors (probably this is a descendent of monarchs-as-patrons); and intellectuals are invited on television to talk about ideas.

We loved working with scholars and tapping into the currents of intellectual thought that make up environmental humanities. I feel a real kinship with the project of EH, the mission of EH, because our work is interdisciplinary too. And I know how hard it is to get people from disciplines to wake up and really talk to each other. It's hard to get busy professionals, working at the top of their game, to slow down and rethink the founding principles of their work, be it scholarly research or artistic inquiry—what I mean by this is to get people to think hard about what "counts" in their work, what makes it real, worthy of respect by colleagues, worth the time it takes to do.

Working on a piece of music theater, I'm trying to make a work that fills the brain with both image and music—but it's hard to get the musicians and composer out of the mindset that "the music needs to be able to stand alone." If it is wonderful music that stands alone, is there really room for the "theater" part? On the flip side, we *use* music in theater all the time, but typically we are asking composers to support, support, support. So for this process I have to live what I preach, and make sure there really is room for the music to be focused on, to fill up the senses, and not always serve the dramatic intentions.

Similarly, it's a challenge for artists to work with scholars be-cause some *ideas* exist better as an essay than on stage, or in a work of art. In my discipline the phrase "that is just an idea" is a kind of insult, a way of saying that "as exciting as this concept may be, it has no dramaturgical teeth, it isn't theatrical." I think that can be hard for scholars to swallow. And it's not a question of interest or importance—it's just a way of communicating that the work of art operates with a different language than the work of scholarship—making some ideas rich with emotion and some details disappear. That's part of the bargain.

I enjoyed so much the presentation at the Timescales conference by philosopher Dale Jamieson, who talked about our sense of moral-ity as both an abstract system of values and something that evolved among great apes living in low-density societies. So the impulse to make art—and to experience it, as audience—I think acknowledges these two modes of thinking, of sensing and being.

BW: Scholars who participated in the etudes you made with us in Phila-delphia, especially at the Timescales conference where the actor/

singers/elders/children performed in a very intimate space, reported that they felt moved to tears—by the dialogues, some with, some without words. Only in more recent years have emotion and affect been a sort-of acceptable part of the scholarly conversation—and not so much in the sciences where a "clinical," or supposedly emotion-free and therefore neutral, mode of practice still most often is the goal. How would you describe the emotional registers you wanted to play on in *PAE*?

DR: Humor, fear, awe, sweetness. Love of humanity. Wonder at the beauty of the universe, patterns of life, existence. We spoke about filmmakers like Kubrick and Malick. My own taste leads me to leaven things with humor, of course.

We also spoke about the uncanny, and the feeling you get when you look at something and *wonder* if it is alive. On the one hand that feeling can verge on horror or disgust—that uncanny valley effect they describe in robotics research. On the other hand, there can be a spiritual feeling, a feeling of connectedness, the feeling that an "élan vital" runs through things that our conscious and scientific minds *know* it cannot. For a long time I thought it was virtuous to do everything I could to suppress this "error" of perception—to be as scientific as possible in all my dealings with the world. But recently I've come to believe that I'm a fleshy monkey like everyone else, with fleshy monkey dreams and emotions, and that it might be more "hygienic" to allow my mind to drift into this illusion of connectedness—that maybe some kinds of understanding might be found only by letting myself, I guess the word is "believe."

So sometimes in the piece we try to evoke the feeling that a celestial body is alive, and sometimes the converse that a mass of people is merely (or wonderfully) a landscape.

I know that we touch on grief with the piece, but more often I spoke about the fact that my feeling in the face of extinctions is not something I have a word for. We come to know death as we grow older—people we love die, we understand the deaths of masses of people in war and disaster, and maybe we understand the extinction of a species like the rhino, and we mourn it. But what are we to think or feel when we contemplate the end of all living things on the planet? What is that feeling called?

At some point in the process, Troy and I talked about "sacred music" and he offered me a thumbnail sketch of how that tradition could be understood: work that evokes "ecstasy; the illusion of time-lessness; and the dissolution of the self." Those emotions or sensa-tions aren't the usual material for theater, and I'm used to getting at them obliquely, or in miniature. But those were the feelings we wanted our audience to approach.

PART II
VARIATIONS, FAST AND SLOW

CHAPTER 7

Time Machines and Timelapse Aesthetics in Anthropocenic Modernism

CHARLES M. TUNG

On the threshold of our various post-Holocene futures, how should we go about representing change over enormous timescales? In his study of the uneven distribution of environmental dangers, Rob Nixon points out that the "slow violence" of ecological disaster is not only occurring "gradually and out of sight . . . dispersed across time and space," but also "playing out across a range of temporal scales." For Nixon, such nonevents unfold out of perceptual range: beyond our "flickering attention spans" and our "spectacle-driven corporate media," whose scope cannot access the "slow paced and open ended" processes that outrun "the tidy closure . . . imposed by the visual orthodoxies of victory and defeat."[1] Timothy Morton has described such nonevents as "hyperobjects"—"things that are massively distributed in time and space relative to humans."[2] In this essay, I would like to reflect on a representational strategy that we do have for registering hyperobjects such as climate change. Timelapse is the dominant aesthetic technique for tracking processes (both large and small) not formatted to the human sensorium or cultural operating system. While we are accustomed to seeing timelapse in eco-cinema as a documentary mode whose objects of concern range from the rise of cities and their rhythms to melting glaciers and deforestation, I would like to connect this kind of large-scale seeing to the speculative mode born at the end of the nineteenth century: the view from the time machine. This way of seeing and its enabling trope travel across certain strands of modernist experiment, ultimately producing a set of questions not only about the nature of time itself, but also about the less philosophical problem of how to gear politics to the range of rhythms, cycles, and spans revealed by modernist time machines.

Modernism in literature and the arts, in the singular, is not usually associated with large, forward-facing timescales. In general, its range of temporal reference, if not located in subjectivity itself, tends to be defined either by backward-facing, "paleomodernist" longings or by small-scale responses to a culture of the short-term, the familiar culture of transitoriness and the moment.[3] However, the speculative element in some early twentieth-century work tacks from Ezra Pound's well-known artistic imperative to "make it new" to the concerns of what Ulrich Beck called "second modernity," reflexively focused (to modify another line from Ezra Pound) on news that will have stayed news longer than we ever imagined.[4] In Beck's reading, whereas modernity was focused on the production and distribution of economic and social goods, second modernity is "the dawning of a *speculative* age in everyday perception and thought," a hyperopic interest in the trajectories and half-lives of "bads" created by modernization, which alters the historicity of the present.[5] While Beck's analyses have been criticized as themselves too narrow a form of sociology, they help us consider a temporal reorientation that Aaron Jaffe has claimed is implicit in modernism from the beginning—a reconceptualization of the historical now in relation to "the future effects of the deep time of modernity."[6] The modern present that has been associated for so long with novelty and brevity in literature and the arts was in fact already being considered in relation to "an unknowable future determinate of the present."[7]

The conjunction of the time machine and certain aspects of modernist time culture can be described as an Anthropocenic modernism. Ultimately I want to suggest that this strain of modernism pluralizes itself via its massive horizontal scans of time that reveal a multiplicity of scales (just as the plural modernisms of our recent global focus are multiplied by horizontal movement across geopolitical spaces).[8] If we start with H. G. Wells's invention of a conceptual and aesthetic device for exploring the far future, many of modernism's textbook examples of canonical temporalities can be understood in a new, second-modernist light—from Virginia Woolf's thoughts of a world "when you're not there" and T. S. Eliot's fascination with trash and the expiry date of tradition, to James Joyce's varieties of scales, Icarian as well as down-to-earth. However, in this essay I want to focus primarily on literature and art that are not usually considered modernist, and I would like to begin with far-futural scope rather than multiplicity. The strain of modernism that includes science fiction—from Wells's *The Time Machine* to Jeremy Shaw's video art *This Transition Will Never End*—is a second modernism of Anthropocenic temporal proportions.

Like Beck's reflexive, speculative age, the epoch of the Anthropocene is the span created by projecting forward. It is characterized by an understanding of the present as the front end of an enormous swathe of time stratigraphically legible for geological periodization tens of millions of years hence. That is to say, the Anthropocene names the long-term impact of "accelerated technological development, rapid growth of the human population, and increased consumption of resources"—those anthropogenic signatures that have been and are being inscribed in the rock layers and ice cores for future scientists.[9] The inhuman aftermath of these textbook features of modernity has been aptly described as a science-fictional scenario by Tim Lenton, a scientist at the University of Leicester: "The alien geologist arriving in the future would see a change in sedimentation rates, novel metals, pollutants, microplastics."[10] Paul Crutzen and Eugene Stoermer, who first proposed the new epoch with an eighteenth-century start date tied to the invention of the steam engine, worked with the Dahlem Conference group to produce a more recent beginning: the sharp rise in certain kinds of human activity (e.g., economic growth, population, energy consumption, telecommunications, transportation, and water use), correlated with the sharp rise in earth-system trends (e.g. carbon dioxide, methane, ocean acidification, terrestrial biosphere degradation) suggests that the alien geologist would see the mid-twentieth century as the origin of long-lasting environmental catastrophe.[11]

We can locate in second modernist cultural production the beginnings of a great acceleration in proleptic geological consciousness and extreme temporal forms (or form-busting temporal extremity), which share with Anthropocene periodization the logic of forward projection, such that the present becomes a stratigraphically significant past from the perspective of the far future. But when we think of the aesthetic and political quandary that Nixon poses—how do we "convert into image and narrative the disasters that are slow moving and long in the making . . . that are anonymous and that star nobody"?—we would be hard-pressed to find anyone proposing modernist literature as an answer.[12] Indeed, the response often goes in the opposite aesthetic direction. For instance, Columbia's Center for Research on Environmental Decisions, a.k.a. CRED, doubles down on downscaled aesthetic forms in order to mobilize political action, recommending that climate science be "actively communicated with appropriate language, metaphor, and analogy; combined with narrative storytelling; made vivid through visual imagery and experiential scenarios."[13] While I do not dismiss the rhetorical effectiveness of CRED's good advice, it is worth considering what modernist experimental techniques, often nonnarrative and even antinarrative in their effects, would contribute to our understanding of slow violence, and how timelapse functions as one of these speculative techniques, even if it seems to be, on one reading, precisely a scaling down to an anthropocentric perspective, a reassertion of "big history" as one smoothly scannable timeline to which all other times can be reduced.

Modernism's use of inhuman prolepsis and the science fiction scenario of the alien geologist are related to one another in their shared impulse to think about large-scale changes over time frames that exceed aesthetic scope and human perspectives. Moreover, beyond just the lengths to which both flash forward, the specifically modernist Anthropocene is an invitation to think across disjunctive multiplicity. The scientist J. B. S. Haldane's creative piece "The Last Judgment" (1927) thematizes scale and disjunction: "it was characteristic of the dwellers on earth that they never looked more than a million years ahead. . . . The continents were remodelled, but human effort was chiefly devoted to the development of personal relationships and to art and music, that is to say, the production of objects, sounds, and patterns of events gratifying to the individual."[14] Haldane's piece features a lecturer on Venus who is looking back from 40 million years away on the destruction of our former home in our despoilment of its fossil (and later, its tidal) energy. Influenced by H. G. Wells's orientation toward the far future and his anxieties about human shortsightedness, Haldane goes 10 million years beyond Wells's famous 30-million-year trip in his *Time Machine,* in order to press his point about foresight in relation to myopic aesthetic scales. For Haldane's Venusian historian, it is aesthetic gratification and its small timescales that were the greatest risk to the planet. But as the teacher scopes out to imagine the apocalyptic end to the planet, the deep prolepsis not only raises the issue (and complicity) of art's default periodicity—moments, days, decades, generations, centuries, and of course the career of the human itself. It also estranges and distances, in modernist fashion, all earthly temporal units. When our need for tidal energy slows the planet's rotation in the year 17,846,151, the fundamental temporal unit of the day "now lasted for forty-eight of the old [twenty-four-hour] days, so that there were only seven and a half days in the year."[15] If nature is taken to be the ground of our understanding of time as an absolute and uniform medium, the temporal scaling up and spatial escape to Venus reveals a variety of clocks keeping various times, and an urgent question about the right clock to live by.

Wells's earlier text also makes temporal scale a matter of plot, and its central trope of the time machine raises the issue of multiplicity as a matter of visibility. The view from the machine appears as a thrilling but also disorienting cinematic effect in its movement through the "palpitations of night and day."[16] As the Traveller accelerates from one second per second to "over a year a minute," and eventually "great strides of a thousand years or more," he notes that "the slowest snail that ever crawled dashed by too fast for me" (77). "I saw trees growing and changing like puffs of vapour, now brown, now green; they grew, spread, shivered, and passed away. I saw huge buildings rise up faint and fair, and pass like dreams. The whole surface of the earth seemed changed—melting and flowing" (76). The text's vehicle produces variable speeds, and its timelapse seems to make its long durations spectacular, vivid, and experiential by compressing them like concertina or accordion bellows, to use Sean Cubitt's metaphor.[17] But as Cubitt points out about contemporary timelapse as data visualization, the "datafication of the photographic image" in fact points us to zones beyond the sensible—governed by the mathematical administration of sense data.[18] In relation to the time machine's effects, the unclarity of the protagonist's speculations in Wells's novel on the social outcomes of late nineteenth-century economic organization and technological developments suggest that, rather than the reduction of all processes down to one smoothly scannable timeline, the machine's most important output is the insight that what comes in and out of view is a function of relative rates—

the speed of, say, a bullet or bicycle wheel in relation to an eyeball or a high-speed camera. To shoot forward in time is to become aware of the relations among various rates, reference frames, and timespaces that are a fundamental condition for successful perception itself. And the leap to the year 802,701 and ultimately to 30 million years in the future makes available not the singular line of human devolution to a blob with tentacles, but rather the insight that the timescape of modern industrial Britain is tangled up with social history, environmental timelines, and evolutionary possibilities.

Early cinema latched onto the view from the time machine without fully tapping the heterochronic potential of this view. Hilariously, the filmmaker Robert Paul was so taken by *The Time Machine* in 1895 that he applied that year for a patent to construct a time-machine simulator out of film, panoramas, and dioramas. As Oliver Gaycken argues, the visual attraction of timelapse, in its "visualization of time scales beyond human perception," can be read as a part of Darwinian discourses and evolutionary cinema.[19] However, the genre of the evolutionary epic and the technique of timelapse in this framework tended to tie the curiosity of the attraction to the sublime. Gaycken cites Ernst Mach as the first to imagine, in 1888, a "telescoping" of time in timelapse, a way of accessing "an experience of deep time": "a quick series of 'magic lantern pictures,'" wrote Mach, could show both the development of the embryo and the species, and "images of a human being from the cradle onwards, thus depicted in his advancing development and then in deterioration into old age in but a few seconds, would have to elicit a . . . grandiose effect."[20]

One could argue that, rather than the grandiose access to inhuman deep time, the technique of timelapse seems to feature a spectacular downscaling and a smooth, frictionless compression. George Pal's 1960 film adaptation brilliantly figures the time machine's timelapse as the easy acceleration of the sun across the chronophotographic cells of the Traveller's solarium window. However, in the 2002 remake of *The Time Machine,* the CGI timelapse sequences and their reliance on erosion algorithms might remind us that the spectacular downscaling and smooth compression of so much data visualization actually dissimulates the disjunctures beneath the suturing, aggregating, interpolating, and averaging of information drawn from instruments of varying sensitivity.[21] If, from a Darwinian perspective,

FIGURE 7.1. The Time Traveller in his machine, his solarium window registering the lapse of days and nights, the decay of his house, and his arrival in the far future. Film stills from George Pal's *The Time Machine*, 1960.

time seems to form the thrilling but stable tracks whose destination was a saleable aesthetic satisfaction and scientific clarity about something beyond human perception, it was also already marked by a heterochrony that animated weirder Einsteinian insights and affects. In *Darwin's Screens: Evolutionary Aesthetics, Time, and Sexual Display in the Cinema*, Barbara Creed explains that early cinema was a direct response to "new and different ways of experiencing time with the development of an array of special effects designed to stop, slow down, and speed up time, to move rapidly from present to past and back again, and to leap from the past and present into the future."[22] In her argument, the science fiction film emerges out of "the strange, uncanny and frightening ramifications of Darwinian theory" and is likewise focused on "morphology or the science of forms."[23] To add to her reading of Pal's "time travel into the future by depicting the accelerated growth of plants and flowers through time-lapse and stop-motion photography," I would emphasize the heterochronic investigation of the morphology of time itself.[24]

The modernist aspect of timelapse that helps us think about the Anthropocene in interesting ways is not the downscaling, and not just the access to a massive, inhuman scale, but the revelation that there are so many scales lurking beyond the range of our senses, even

FIGURE 7.2. Erosion algorithms in the timelapse sequence of Simon Wells's *The Time Machine,* 2002. *The Time Machine* copyright 2002 by DW Studios, L.L.C. All rights reserved.

as they overlap in, cut across, and constitute "the" present. As Keith Williams says of Wells's text in relation to cinema and modernity, "Time travelling . . . is endemic to both high and low modernism, but seems to derive from a kind of Wellsian relativisation (accelerated, dilated, reversed, subjectivised) towards the turn of the century."[25] Williams reminds us that Wells's serially published text in the *National Observer* (March–June 1894) emphasizes the "relativity of the in/visible": the Traveller remarks that "as soon as the pace became considerable, the apparent velocity of people became so excessively great that I could no more see them than a man can see a cannon-ball flying through the air."[26] If we think about what happens in modernist fast-forwards, smaller-scale temporalities do not always dissolve into a largest, most fundamental one. While the thought of stars in Woolf's *Night and Day* (1919), for example, "froze to cinders the whole of our short human history, and reduced the human body to an ape-like furry form, crouching amid the brushwood of a barbarous clod of mud," the jump cut in *Mrs. Dalloway*—from shopping and traffic in London in 1923, to an overgrown future littered with bones, wedding rings, and decayed teeth—demonstrates a temporal hyperopia in which a variety of timescales (the myriad rhythms of social strata, biopolitical management of different walks of life) are not flattened or elided but rather implicated in inhuman timescapes.[27] Olaf Stapledon's strange text *Last and First Men* (1930), which stretches the form of the novel over two billion years of human history, puts an incredible amount of proleptic pressure on a semicolon in his "Time

Scale 5"—"Planets formed; End of Man." Each of Stapledon's five time scales seem to demonstrate the way the history of humans, even in a two-billion-year fantasy, is dwarfed by successively larger timelines. In plot terms, the novel fills in the between-time with a march of episodes of human evolution, but the text's mental-time-travelling narrator courts a madness mirrored not just in the story's scope and sweep but in the fact that he "might behold the events of a month, or even a lifetime, fantastically accelerated so as to occupy a trance of no more than a day's duration."[28] The variability and multiplicity are as threatening to narrative and psychological coherence as the compression.

The engineer and inventor Danny Hillis, while he was still a VP at Walt Disney Imagineering, dreamed up a temporal scaling device that is now called "The Clock of the Long Now," which he imagined would be capable of running for 10,000 years and of inspiring foresight. As Stewart Brand recounts in his book on the inception, construction, and significance of the project, Hillis also sees the Clock of the Long Now as a kind of theme-park ride: "Time is a ride," Hillis has said, "and you are on it."[29] But the view from the time machine reveals this ride to be a much stranger trip than Hillis and Brand conceive of. The fantasy of riding the time machine reveals, in its usefully ungratifying ellipses and extreme prolepses, the disjunctive intersection of rates, scales, lines, and reference frames. Hillis's and Brand's Anthropocene attraction projects the visitor as a future human, who, after walking through the various rooms of the gigantic

FIGURE 7.3. *(left)* Model of the Clock of the Long Now at the Long Now Foundation office in San Francisco and *(right)* the giant pendulum of the 10,000 Year Clock being built into a mountain in Texas. Photograph courtesy of the Long Now Foundation.

clock-built-into-the-mountain, discovers that human survival had depended on "the people of this ancient time [who] had the foresight to think this far into their future and create this place."[30] The visitors of this grand ride affirm the present as the guarantee of historical continuity for the next 10,000 years by becoming their own descendants delighting in the wisdom of thinking the ultimate long now. By contrast, Woolf's "curious antiquaries" in *Mrs. Dalloway*, "sifting the ruins of time" in an overgrown future London, and alien geologists are figures of modernist speculation: they help us think beyond the inevitability of death or the imperative of survival on existential and historical scales, to the various distended timescales necessary for thinking the future effects of the present—effects of imperialist synchronization, capitalist short-termism, and the uneven distribution of heterogeneous, long-lasting trajectories of harm.

In her diary entry for September 20, 1920, Woolf recounts a conversation with Eliot about his fragmentary style in which she "taxed him with wilfully concealing his transitions," and it has been tempting to think about Woolf's corner of modernist experiment as characterized perhaps by the ontologization of an undifferentiated, all-encompassing, Bergsonian process of transition.[31] In Henri Bergson's philosophy, time is a flow in which changes "melt into and permeate one another, without precise outlines, without any tendency to externalize themselves in relation to one another, without any affiliation with number."[32] But in my view, modernism's most interesting en-

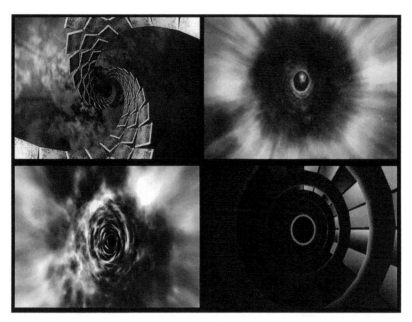

FIGURE 7.4. Stills of wormhole travel from Jeremy Shaw's video montage *This Transition Will Never End #4*, 2008. Courtesy of Jeremy Shaw and KÖNIG GALERIE, Berlin.

gagement with change over time is best teased out by connecting it to a more contemporary, scaled-up exploration of timelapse. In Jeremy Shaw's video art series *This Transition Will Never End*, for instance, begun in 2008 with plans for never-ending additions and updates, temporal process is figured as or by a video montage of wormhole clips from various time-travel TV shows and films—"appropriated footage of the un-representable," as Lars Bang Larsen puts it.[33] It is, in Shaw's words, an "ongoing archive of appropriated footage taken from a wide variety of movies and television in which a vortex, or any such tunnel-like or spiraling image, is used to represent the slippage of time or a transition from one reality to another."[34] As Shaw remarks, these vortical representations of dimensional portals figure the history of moving pictures as a certain kind of temporal special effect, while also highlighting the psychedelic fantasy of liberation from everyday constraints and coordinates as "a ubiquitous representational cliché of the transition into another reality, state, time, etc." that is paradoxically "visually un-documentable."[35]

Shaw's collection of clips offers a critique of the fantasy of access to hyperspace—what David Wittenberg's book *Time Travel: A Popular Philosophy of Narrative* identifies as the narratologically necessary "viewpoint-over-histories."[36] But the jerky ride of the mash-up also picks up on the disjunctive aspects of modernist style, refusing the concept of the total break within a singular timeline: as he suggests, the fantasy of time travel, like the psychedelic trip, is "not about 'dropping out' to the utopian community or ideals of yesteryear, but it's an acknowledged engagement and dialogue with this proposition of time-slippage."[37] Larsen summarizes the wormhole excursion in a way that is useful for thinking timelapse as an Anthropocenic modernist technique:

> Dropping out can also be a sidestepping of the present, meaning a speculation about a different concept of time and what we do not yet know. This aspect of futurity is very important for psychedelia as a critical art form—a criticality that isn't moralistic but is at the same time excessive and speculative and has a dimension of madness too.

The discussion of effects also points to the contrast between phenomenological-type psychedelic works that refer to a supposed fullness "behind" or "before" the representation, and the idea of another temporality—an *Aion*—where it isn't that simple to divide things up in pre- or post-.[38] The deep timelapse over immense scales has a history in the modernist time machine, a vehicle driven by a distinctly heterochronic Aion. In *How to Build a Time Machine*, the physicist Paul Davies says that if one were to observe the Earth from a neutron star, where time is elapsing 30 percent more slowly, we would see "terrestrial events speeded up, like a fast-forward video show."[39] In second modernist fantasies, the position of viewpoint-over-histories is not the godly space of eternity, but simply the location within a different reference frame and timespace that is running at a different rate. What the Anthropocene asks us to do is not simply to extrapolate forward but to think through the tensions among scales in relation to our long-term future. As Dipesh Chakrabarty's now well-known query into the intersection of postcolonial studies and climate change points out, our challenge today is to "think human agency over multiple and incommensurable scales at once," to engage "the necessity of thinking disjunctively."[40] In relation to multiple kinds of agency, timelapse, at least in its Anthropocenic modernist operations and in its aesthetic failures to provide coherent satisfactions formatted for humans, is a way of thinking scalar misalignments and overlap, in which both catastrophe and possibility are written.

NOTES

Many thanks to Bethany Wiggin, Carolyn Fornoff, and Patricia Kim for creating an exciting exchange of ideas at Timescales: A Conference of the Penn Program in Environmental Humanities, University of Pennsylvania, October 20–22, 2016.

1. Rob Nixon, *Slow Violence and the Environmentalism of the Poor* (Cambridge, Mass.: Harvard University Press, 2011), 2–3, 6.

2. Timothy Morton, *Hyperobjects: Philosophy and Ecology after the End of the World* (Minneapolis: University of Minnesota Press, 2013), 1.

3. I give an account of this underexplored aspect of modernist aesthetics in "Baddest Modernism: The Scales and Lines of Inhuman Time," *Modernism/modernity* 23, no. 3 (2016): 515–38.

4. I take this witty formulation from Mark McGurl, who describes how the warning signs that the U.S. Department of Energy envisioned could steer future humans away from a hazardous nuclear waste facility in New Mexico: "Like literature as conceived by Ezra Pound, the information conveyed by [the storage facility's] marker will remain current for an extremely long time—it will be 'news that STAYS news.'" Mark McGurl, "Ordinary Doom: Literary Studies in the Waste Land of the Present," *New Literary History* 41, no. 2 (2010): 330.

5. Ulrich Beck, *Risk Society: Towards a New Modernity,* trans. Mark Ritter (London: Sage, 1992), 73.

6. Aaron Jaffe, *The Way Things Go: An Essay on the Matter of Second Modernism* (Minneapolis: University of Minnesota Press, 2014), 19.

7. Jaffe, 19.

8. Utopian fictions have a critical history that of course begins much earlier than the late nineteenth and early twentieth century. It is important to note that modernist heterochrony departs significantly from visions of the u-topos and heterotopia figured as destinations of an altered singular history, negations of place in the imagination of other spaces that are sometimes arrived at instantaneously in sleep time travel. Scaled-up heterochronic scans focus not only on the end result of a divaricating sequence but on the variety of histories, rhythms, and scales lurking in what we take to be the ultimate historical line.

9. Colin N. Waters et al., "The Anthropocene Is Functionally and Stratigraphically Distinct from the Holocene," *Science* 351, no. 6269 (January 8, 2016): 2.

10. Tim Lenton, quoted in Tom Chivers, "The Anthropocene Age: What

World Will Humans Leave Behind?" October 17, 2014, http://www.telegraph
.co.uk/news/science/11167165/Scientists-wonder-what-in-the-world-will-we
-leave-behind.html.

11. See Will Steffen et al., "The Trajectory of the Anthropocene: The Great
Acceleration," *The Anthropocene Review 2*, no. 1 (April 1, 2015): 81–98.

12. Nixon, *Slow Violence and the Environmentalism of the Poor*, 3.

13. D. Shome et al., *The Psychology of Climate Change Communication: A Guide
for Scientists, Journalists, Educators, Political Aides, and the Interested Public* (New
York: Center for Research on Environmental Decisions, 2009), 2, http://guide
.cred.columbia.edu/index.html. http://cred.columbia.edu/about-cred/.

14. J. B. S. Haldane, "The Last Judgment," in *Possible Worlds* (London: Chatto
and Windus, 1927; repr., New Brunswick: Transaction Publishers, 2002), 295.

15. Haldane, 298.

16. H. G. Wells, *The Time Machine: An Invention*, ed. Nicholas Ruddick (Peter-
borough, Ont.: Broadview Press, 2001), 77.

17. Sean Cubitt, "Everyone Knows This Is Nowhere: Data Visualization and
Ecocriticism," in *Ecocinema Theory and Practice*, ed. Stephen Rust, Salma Monani,
and Sean Cubitt (New York: Routledge, 2013), 282.

18. Cubitt, 282

19. Oliver Gaycken, "Early Cinema and Evolution," in *Evolution and Victorian
Culture*, ed. Bernard V. Lightman and Bennett Zon (Cambridge: Cambridge University Press, 2014), 97.

20. Ernst Mach, "Bemerkungen über wissenschaftliche Anwendungen der
Photographie," *Jahrbuch für Photographie und Reproductionstechnik 2* (1888),
284–86, quoted in Gaycken, "Early Cinema and Evolution," 108.

21. See Cubitt's reading of the way big data is crunched in climate analysis in
"Everyone Knows This Is Nowhere," 283.

22. Barbara Creed, *Darwin's Screens: Evolutionary Aesthetics, Time, and Sexual
Display in the Cinema* (Carlton, Vic.: Melbourne University Publishing, 2009),
xix.

23. Creed, 39–40.

24. Creed, 57. I give an account of heterochrony in evolutionary developmental biology and modernist aesthetics in "Modernist Heterochrony, Evolutionary
Biology, and the Chimera of Time," in *The Year's Work in the Oddball Archive*, ed.
Jonathan P. Eburne and Judith Roof (Bloomington: Indiana University Press,
2016).

25. Keith Williams, *H. G. Wells, Modernity, and the Movies* (Liverpool, U.K.:
Liverpool University Press, 2007), 3.

26. Williams, 10; H. G. Wells, *The Definitive Time Machine: A Critical Edition
of H. G. Well's Scientific Romance*, ed. Harry M. Geduld (Bloomington: Indiana
University Press, 1987), 158.

27. Virginia Woolf, *Night and Day* (London: Duckworth Press, 1919; repr., New

York: Barnes & Noble Classics, 2005), 172; Virginia Woolf, *The Mrs. Dalloway Reader* (Orlando: Harcourt, 2003), 207.

28. Olaf Stapledon, *Last and First Men* (London: Methuen, 1930; repr., Mineola: Dover Publications, 2008), 180.

29. Stewart Brand, *The Clock of the Long Now: Time and Responsibility* (New York: Basic Books, 1999), 69.

30. Brand, 47.

31. Virginia Woolf and Anne Olivier Bell, *The Diary of Virginia Woolf, Vol. 2, 1920–1924* (New York: Harcourt Brace Jovanovich, 1978), 67.

32. Henri Bergson, *Time and Free Will: An Essay on the Immediate Data of Consciousness* (1889; London: G. Allen, Limited, 1913), 104.

33. Lars Larsen, "PCP: Pop/Conceptual/Psychedelic: Interview with Jeremy Shaw," *C Magazine*, no. 101 (2009): 19.

34. See "Jeremy Shaw—Degenerative Imaging in the Dark | Lambda-LambdaLambda | Artsy," accessed May 15, 2017, https://www.artsy.net/show/lambdalambdalambda-jeremy-shaw-degenerative-imaging-in-the-dark.

35. Larsen, "PCP," 20.

36. David Wittenberg, *Time Travel: The Popular Philosophy of Narrative* (New York: Fordham University Press, 2013), 131–47, as well as chap. 5, 148–77.

37. Larsen, "PCP," 21.

38. Larsen, 20.

39. Paul Davies, *How to Build a Time Machine* (New York: Penguin, 2002), 21.

40. Dipesh Chakrabarty, "Postcolonial Studies and the Challenge of Climate Change," *New Literary History* 43, no. 1 (2012): 1–2.

CHAPTER 8

Fishing for the Anthropocene
Time in Ocean Governance
JENNIFER E. TELESCA

Goethe's Hourglass: Making the Anthropocene

Sojourning in Venice on his passage through Italy, the German writer and statesman Johann Wolfgang von Goethe recounted "the public discussion of a law case" he witnessed in the "spacious hall" of Palazzo Ducale, once the center of the Venetian Republic. In his diary dated 3 October 1786, Goethe characterized the proceeding as a "comedy." Amid the tedium interrupted by "prepared" jokes, "the judges knew what they had to say, and the parties what they had to expect." Even so, the scripted performance was "completely crammed" with onlookers "boasting" of their princess—the wife of the Duke—who seemed not above the law. She was obliged to appear before the court, "in her own palace," so that nobles could assert their right to recover inherited property from her estate. Goethe observed of this ritual act in high society centuries ago, which highlights law's temporal dimensions:

> When the clerk began to read, I for the first time clearly discerned the business of a little man who sat on a low stool behind a small table opposite the judges . . . More especially I learned the use of an hourglass, which was placed before him. As long as the clerk reads, time is not heeded, but . . . As soon as the advocate opens his mouth, the [hour]glass is raised, and sinks again, as soon as he is silent.[1]

For Goethe, the "little man" who kept time by hourglass was not a trivial prop in a state drama. The legal anthropologist Carol Greenhouse remarks, referencing Goethe's memoir: "The bureaucrat with the hourglass did not just keep the time; he *started* and *stopped* it. He literally *created* time and *ended* it with each intervention."[2] Turned this way or that, laid on its side or held upright, the hourglass was a core technology in a courtroom spectacle that expressed the power of the state to secure wealth for the privileged few by asserting title over space and, significantly, over time.

Goethe's chronicle of probate in late eighteenth-century Venice anticipates this chapter's interest in ocean governance, emphasizing the ways bureaucrats have put technologies of time to work to affirm their authority over wealth derived from commercial fish on the high seas. By drawing on ethnographic vignettes of my own—based on my experience as an accredited observer of the International Commission for the Conservation of Atlantic Tunas (ICCAT)—this chapter borrows anthropological insights to show that marine policymakers have not occupied a time-space already made for them. On the contrary, marine policymakers have contributed determinatively to the making of the time-space they occupy: the Anthropocene. They have accomplished this, I argue, by using such tools as visual charts, scientific models, and statistical formulas that together serve as reference points to plan, measure, and quantify time as an exercise of power.[3]

This point is more than academic. Given the legalized slaughter of commercial fish worldwide since World War II, this chapter shows that the technocratic establishment must be seen as a collaborative agent in accelerating the planetary development of the Anthropocene. That policymakers have, since the mid-twentieth century, extracted fish on a scale and at a pace never before known on Earth is not an unintended byproduct of a technocratic way of life but one of its vital preconditions. The time-based assumptions inherent in the mathematical principle of maximum sustainable yield (MSY)—first codified in international law in the 1950s—established the conditions for maritime nations to haul countless creatures out of the sea for their calculable gain. Despite dreadful results, MSY is still the technocrat's most trusted tool of authority no matter the fishery the world over.

The task of this chapter is to illuminate how—or by what time-specific logics, mechanisms, and modes of account—technocrats have realized the unremitting destruction of the ocean and the creatures living there since the 1950s. While scholars rightly debate the enormous implications of designating the Anthropocene as the new epoch of geological time, this chapter emphasizes instead the importance of another moment—the middle of the twentieth century—

when geologists say the Anthropocene swung into another stage: the Great Acceleration. The Working Group on the Anthropocene, part of the International Union of Geological Science, regards the Great Acceleration as a time when carbon emissions, nuclear waste, plastic pollution, cement, and bones from the domesticated chicken industry spiked across the globe with such rapidity after World War II that planetary life profoundly changed. This development doubled down on the prospect for the sixth mass extinction, the first mass extinction event in the planet's four and a half billion-year history to be caused by one species—that is, our own.[4]

"Maximizing" "Yield" in the Great Acceleration

It is not by coincidence that the rise of the Anthropocene's Great Acceleration marks the advent of the contemporary architecture for ocean governance.[5] One of its aims was to constrain—so as to enable the prolongation of—the "ruthless efficiency"[6] of fish extraction already underway. An important development came in 1955 when the International Technical Conference on the Conservation of the Living Resources of the Sea met in Rome in anticipation of the four treaties concluded in the UN Conference on the Law of the Sea (UNCLOS) in 1958.

At the Rome Conference, nation states agreed, by a one-vote margin,[7] to base fisheries policy on the statistically derived principle of maximum sustainable yield (MSY). This juridical formula offers a scientific rationale for fleets to fish as hard as possible, not only to feed people but also to generate profits for a country's economic growth. To regulate exports in commercial fish in domestic waters and on the high seas, the technocrats locked the animals in time by allowing them to be possessed and traded right up to the threshold of collapse. Now the foundation for fisheries policy worldwide, MSY had as its greatest champions diplomats from the United States who at the dawn of the Cold War aspired for their own commodity empire in the relentless promotion of capitalism's supremacy over communism.[8]

Formed on the heels of UNCLOS in the 1960s, the treaty body that is the subject of this chapter—ICCAT—replicates in miniature a representation of time corresponding with what Pope Francis calls the "technocratic paradigm."[9] ICCAT promotes a "regime of value"[10] that assumes the treatment of the nonhuman as "animal capital" is justifiable, as if fish are above all else objects to be mastered, manipulated, conquered, and possessed for the expansion of trade.[11] Conserved, in this worldview, are not the fish per se but the export markets of well-financed ICCAT member states and the power derived from them on the world stage.

The calculative ethos on which this worldview depends did not just happen in time. The social theorist Pierre Bourdieu writes: "An economic organization which tends to ensure predictability and calculability demands a particular disposition toward time and, more precisely, *toward the future*, since the 'rationalization' of economic conduct implies that the whole of existence be organized in relation to an absent, imaginary vanishing point."[12] An emphasis on time cast in the future corresponds with technocratic practices in capitalist modes of economy. To calculate and predict economic growth, time takes the character of the linear and becomes a site of instrumentalization, rationalization, and commodification over which the technocrat attempts to assert control, the subject to which I now turn.

Anthropological Reckonings: On Linear and Cyclical Time

Before I offer ethnographic descriptions of technocratic time, it is worth emphasizing that time is not a singular, universal experience or a "homogenous medium within which something like an economy or society unfold[s]."[13] Peoples the world over, through millennia, have related to time in different ways: a fall, a return, a continuous presence, a noncumulative stasis, a measured duration.[14] In her study of law, Greenhouse renders time social according to two formal states, which are not mutually exclusive: the linear, discussed below, and the cyclical, which emphasizes continuous repetition between opposed, reciprocal domains, such as life and death, youth and age.[15]

Key for the purposes of this chapter is the suggestion that life does not simply happen *in* or *through* time-space, as the anthropologist Nancy Munn made clear long ago in her study of Kula exchange in Papua New Guinea. She writes, "Sociocultural action[s] . . . do not simply go on *in* or *through* time and space, but . . . they form (structure) and constitute (create) the spacetime . . . in which they 'go on.'"[16] More simply, people do not inhabit time; they make the time they inhabit. This decades-old anthropological insight acquires new meaning given the temporal demands of the Anthropocene.

To be clear, this chapter does not consider time a social construct let loose in the cosmos. Instead it emphasizes that the cultural representation of time operating in diplomatic circles is hinged to and becomes "a medium for the achievement of hierarchical power and governance."[17] The exercise of power over time is a way to govern beings, including the nonhuman animals ICCAT attempts to master and control. Time at ICCAT is thus not natural, let alone rooted in Nature, but constitutes "a set of cultural claims about the efficacy of law and specific technologies of social ordering."[18] Time only seems about Nature—the seasons pass, the moon waxes and wanes—because Western cultures tend to assume or take for granted aspects of social time that are so deeply embedded they appear self-legitimating and natural in their formation.

In the main, time at ICCAT is experienced as linear, unidirectional, as if progressing forward, without reverse, characterized by a start and an end, a cycle broken, full stop. A geometric connection to time draws on the Judeo-Christian tradition wherein the "origin of time in creation and . . . the end of time in a day of judgment" generates an ordinal image or fixed sequence of past, present, and future.[19] Popularized since at least the rise of capitalism and print technologies, linear time in the Anglo-European West "passes as if it were really abolishing the past behind it."[20] Faithful to the idea of "progress," as if technocrats perfect a flawed or backward past, the ICCAT regime finds its temporal base here. As an institution of high modernity, ICCAT is not ensnared in traditions gone by but instead directs its energy forward, as if harnessing a future bound up in uncertainties about sea creatures disconnected from webs of life.

Linear time is, for Greenhouse, time with a purpose. It does not displace other experiences of time, including the cyclical, as much as it dominates them. Institutional settings such as ICCAT make linear time relevant. That linear time dominates technocratic life is an expression of its efficacy in constructing and perpetuating social institutions that are themselves hierarchically ordered. The church that sought conversion among its clergy, the monarchs that engaged in building the nation-state, the propertied elites that found in the image of "progress" their own symbols of legitimacy—these are the ways in which power and technologies of linear time are connected.[21] "The fact that debates over wages, interest payments, and secular government in general took place" according to the ticking clock whose time slipped and evaporated "underscores the extent to which linear time calls forth" and declares, ultimately, time instrumental in social terms.[22]

Let me now establish the common ways in which linear time is realized in a high modernist, secular institution that must convene year after year to maintain and affirm its regulatory control over sea creatures[23] "against a host of forces that make legitimacy provisional," incomplete, insecure.[24] It is precisely in such ritual performances as an ICCAT Commission meeting that linear time crops up and reigns.[25]

Linear Time in Action: On the "Future" of "Uncertain" "Resources"

In November 2011, on the still currents of the Marmara Sea, delegates representing state, market, and environmental interests convened for the annual Conference of the Parties to ICCAT, known for decades as, pejoratively, the "International Conspiracy to Catch All Tunas."[26] In Pendik, a suburban district of Istanbul, some fifty member states[27] met behind closed doors to settle the fate of the remaining fish on the high seas of the Atlantic Ocean and its nearby seas, such as the Mediterranean: yellowfin, albacore, and mako shark, to name three. Gross domestic products, trade deficits, jobs in the maritime sector, the illegal traffic of goods, even fealty to the nation-state as a sea power were all at stake as delegates decided how to regulate the catch of itinerant sea creatures killed to supply global markets. ICCAT delegates numbered nearly five hundred that year. This ethnographer was one of them.[28]

As winter approached those gray, damp days at the posh Green Park Pendik Hotel and Convention Center, I crossed paths with a diplomat on a morning coffee break.[29] Hair carefully coiffed, physique trim and tall, he dressed in a smart suit colored midnight blue with a yellow tie that popped and accentuated his demeanor. More like a dapper CEO of a multinational corporation than an unkempt, stale bureaucrat pushing paper, he embodied the propertied class symbolized in Goethe's hourglass. He made time.

Beside potted ginkgo trees in the grand foyer of polished marble mirroring passersby—as if exhibiting the "progress" the emerging market of Turkey had achieved in secular modernity—the diplomat reflected, gesturing to the crowd of ICCAT attendees: "Delegates here are concerned about the future. It's uncertain. We have to think ten, fifteen, twenty years ahead . . . Who will control these resources?" He speculated on how his BRICS (Brazil, Russia, India, China, South Africa) country would compete for finite "resources" as rich member states in Europe, Japan, and the United States jockeyed for pole position through international organizations such as ICCAT. Implicit in his words was the understanding that the thirst for empire over the centuries structured ICCAT's race to fish until the ocean bottomed out, even though the official rules codified in the ICCAT Convention made no mention of history. ICCAT must erase the past to give the appearance of fair play and maintain the fiction that all member states, including former colonies, volunteered to participate in international law as equals.[30]

It is worth reiterating that the diplomat characterized time in this space of high modernity as "uncertain," decidedly forward looking, unencumbered by the past, narrow, shallow, brief, and measured in decadal increments of years instead of centuries, let alone geological epochs such as the Anthropocene. The diplomat spoke as if the nonhuman animals under ICCAT's remit were outside of history, everlasting regardless of fishing pressure, unconnected to human societies in intimate and time-specific ways.[31] Linear time was ascendant. History disappeared. The past only flashed as remnants in documents rife with mystifying techno-speak, sanitized of politics. ICCAT member states bring this world of linear time into being by consenting to rules that take as their primary ground of authority fisheries science and its probabilistic forecasts about the future of assets in ICCAT's portfolio—better known as fish "stocks."[32]

The Matrix

When the ICCAT treaty entered into force in 1969, the Commission formed to carry out its mandate had declared in its founding text its primary objective: "maximum sustainable catch." The technocrat's special device for overseeing commercial fish—maximum sustainable yield (MSY)—was by now widespread. Proponents of MSY assumed that statistics, equations, and mathematical models would help fisheries management do its job: to find that sweet spot between catching too little and catching too much by drafting regulations that prevented commercial catch from exceeding its limit.

Alienated from their caretakers, commercial fish became mere cogs in the machine, their size and age ignored, as if sea creatures were automatons who were supposed to know, conform to, and obey the ways in which the experts organized them—based on how they bred and interacted with one another—so as not to upset the "yield."[33] MSY was both a quantitative estimate of, and a tool used to extract maximum profit from, the animal world. It relied on numerical time series, which sequenced successive points to evaluate a cyclical crisis, or whether enough fish were born to replace the ones that died. Linear and cyclical times were not so much distinct as disaggregated, the former aiming to contain and surmount the uncertainties arising from the latter. Conservation biologists have long critiqued MSY for these and other reasons.[34]

Statisticians calculated MSY during regular checks on the volatility of fish "stocks" in "assessments" by scientific committee, just as financial analysts on Wall Street review the potential for returns on investment.[35] ICCAT member states determined the frequency of "stock assessments," which happened every few years for only the most valuable of commercial fish, such as swordfish. That the statisticians were asked to evaluate variation in compressed intervals deeply frustrated the ones I met. Two years here, three years there did not allow the statisticians time to accumulate enough data points to show policymakers whether a change in forecast was significant. Meanwhile, fisheries managers were eager to know whether and by what magnitude they could adjust their targets. The whole enterprise depended on these fabricated moments when scientists issued their revised MSYs in "stock assessments" so that fisheries managers could (re)negotiate the total allowable catch (TAC), sliced up as quota for some member states. How many tons of, say, bigeye tuna may Chinese Taipei (Taiwan) legally take from the Atlantic Ocean next year?

This was complicated business. Managers without training in fisheries science complained about how dense the "stock assessment" reports were. Experts created the Kobe Matrix to simplify matters.

Developed at a meeting of regional fisheries management organizations (RFMOs) in Kobe, Japan, in 2007, the Kobe Matrix is a graphic depiction of MSY with a twist. It plots the modeled outputs that forecast uncertainties in "stock" based on calculations of MSY—once affixed to probability estimates. The Kobe Matrix divides a two-dimensional plane into four quadrants of right angles by color: a red quadrant signifies danger or the likelihood of overfishing, yellow caution, and green go. When point estimates about MSY cluster in one of the quadrants, fisheries managers are supposed to adopt policy accordingly. Policymakers anticipate the reference points crowding the red quadrant, which signals overexploitation and the need to pull back on "harvest."[36] The prospect of achieving an outcome absent overexploitation can be adjusted to different success rates at, say, 75 or 90 percent, once applied to different futures. What is the likelihood at a probability of 60 percent—the dominant increment ICCAT used—that overfishing Atlantic bluefin tuna would occur in five, ten, twenty years if *this* MSY is adopted? The lower the probability adopted, the less likely the appearance of overexploitation and, with it, policy failure.

An elaborate statistical method developed to extract fish in complex capitalist economies, the Kobe Matrix speaks of manifold futures in linear time. As the future of fish forks into multiple probabilities, volatility shifts as the object changes according to inputs in the forecasting models. Like traders swapping currency, fisheries managers bet on the future performance of an underlying entity—a fish "stock" in a futures market—with the delivery date those years cast in the probabilities of the Kobe Matrix. The sociologist Randy Martin writes, "The techniques and logic of risk management, of turning the uncertain into a probability with measurable outcome and calculable gain, roam" through institutions such as ICCAT.[37] They bring the future into the present—which, says Martin, is the temporality of financialized risk attuned to profitable speculation.[38]

Back at the ICCAT Commission meeting, member states adopted Resolution 11–14 without fanfare in that swank hotel in Istanbul. Its aim: "to standardize the presentation of scientific information" emanating from ICCAT's scientific committees.[39] Guidelines on how to adopt the Kobe strategy occupy the bulk of this ICCAT Resolution. Although not binding on ICCAT member states, the Resolution assured policymakers that the charts, graphs, scoring tables, rubrics, and grids produced by ICCAT scientists would be uniform, tidy, free of clutter in the executive reports organized by "stock." It encouraged scientists to more swiftly translate dense, specialist knowledge to lay fisheries managers on-the-go and at-the-ready during this ten-day ICCAT Commission meeting when time was tight and exports dear.

The Resolution's economistic motivation was clear: reduce the time fisheries managers spent reviewing the lengthy, tedious files to achieve bureaucratic "efficiency," since the parameters for maximum catch were immediately known by the visual graph of the Kobe Matrix. Bruno Latour writes: *"It is the sorting that makes the times, not the times that make the sorting."*[40] The Kobe Matrix as a graphic representation of MSY did not simply forecast volatility in the future supply of commercial fish as much as it provided the protocols by which speculation operated.[41] The Resolution adopted in Istanbul did not challenge the authority of MSY as the dominant tool used to develop fisheries policy for decades. Instead it intensified the mathematical model's capacity to simplify complex phenomena on the high seas through the Kobe Matrix. Change in mathematical representations did not necessarily imply better policy outcomes or less overfishing. The emperor has new clothes.

As the ocean continued to appear "as a two-dimensional, air-sea interface"[42]—as a zone for vessel operations and as a means for "geoeconomic"[43] calculation—the Kobe Matrix has become a device that (re)projects and secures linear time against a host of forces that threaten technocratic power over maritime space. It collapsed futures and yesteryears into one flat, convenient plane, and reentrenched graphically the rationale that fish were first and foremost a mutable—and adjustable—inventory purposed for capitalist extraction in the realm of commodity empires.[44] The "geometry" of time represented in the lines, plots, and pie charts of the Kobe Matrix rendered official—and yet still buried—Western assumptions that "link the meanings of space to dominion," in the words of Greenhouse.[45] "The difficulty with such ethno-centrisms is that they export to other cultures a mystification essential to modern Western political thought, which is that the meanings of time exist apart from the logics of power and accountability."[46]

The authority of fisheries management can no longer be separated from time-space given the sharp rise in overexploited fish the world over during the Anthropocene's Great Acceleration. In 2018, the most recent year for which data are available, fleets have overexploited more than one third of global fish "stocks" beyond their biologically sustainable limit, while the world's appetite for fish and fish products shows no sign of abatement.[47] Consider that these statistics—provided to the UN's Food and Agriculture Organization (FAO) by nation-states—are conservative estimates.[48]

The Future Is Now: Time and the Annihilation of Oceanic Space

At an Ottoman palace overlooking the Bosporus, ICCAT's annual gala dinner in 2011 featured a performance by whirling dervishes under crystal chandeliers. Off the clock, free from official duty, without meetings in session, linear time receded as the delegates reveled in the host country's traditional past. Time seemed suspended in the dervishes' vortex of bodily movement, which held the delegates in trance quite unlike the DJ after dinner with his pop electronic sound. ICCAT delegates consumed the spectacle as if it were commercialized theater, unable to access on their own the Sufi remembrance to God practiced by the dervishes since the thirteenth century. Evacuated of the mystical, the ritual that once marked a spiritual ascent unfettered by material trappings could not escape its context as mere entertainment for these high modernizers. Regimes such as ICCAT produced both the time of the Great Acceleration and the disposition of the bureaucrats who acted within and through this world.[49]

The "uncertain becoming" of high-seas fish, "strewn with tipping points" in the Anthropocene, "scarcely resembles the radiant future promised" by the technocrats when ICCAT formed a half century ago.[50] The regime's power to kill by quota—conceived according to the "continuous progress unfurling to the rhythm of productivity gains"[51]—must be understood as "part of the hermeneutics of the state, which claims the monopoly over legitimate violence."[52] The technocrats that made the Great Acceleration through regimes such as ICCAT legalized a kind of death spiral by sanctioning the rapid slaughter of sea creatures using tools such as MSY, now wrapped in the graphic simplicity of the Kobe Matrix.

Seen from the perspective of the Anthropocene, fisheries management as currently constituted enacts the violence of managed extinction and represents the disavowal of the sanctity of life. It reduces nonhuman animals to number, to dots plotted on graphs, to calculable biological assets full of profit potential. To expose this destructive logic and routine way of being-in-the-world is to begin to find a place outside it from which to know the world differently or to resist directly the brutal effects of ICCAT's own engineering.[53]

Whether the master's tools will help to dismantle the master's house remains to be seen.[54] Meanwhile, the technocratic establishment continues to stalk this Earth.[55] It ignores if not denies its role in accelerating the Anthropocene through financialized forms of risk management under extractive capitalism, which has annihilated oceanic space by linear time.[56]

NOTES

My thanks to Matthew Canfield, Peter Dimock, Ann Holder, Johanna Römer, Robert Wosnitzer, the editors of this volume, and anonymous reviewers for comments on earlier drafts.

1. Johann Wolfgang von Goethe, *Goethe's Travels in Italy* (London: George Bell and Sons, 1885), 64–66; available online at https://warburg.sas.ac.uk/pdf/ndn46ob2788755.pdf (accessed January 2, 2018).

2. Carol J. Greenhouse, *A Moment's Notice: Time Politics across Culture* (Ithaca, N.Y.: Cornell University Press, 1996), 22 (emphasis in original).

3. Louis L. Bucciarelli, "Engineering Design Process," in *Making Time: Ethnographies of High-Technology Organizations,* ed. Frank A. Dubinskas (Philadelphia: Temple University Press, 1988), and Nancy D. Munn, "The Cultural Anthropology of Time: A Critical Essay," *Annual Review of Anthropology* 21 (1992): 104.

4. Elizabeth Kolbert, *The Sixth Extinction: An Unnatural History* (New York: Henry Holt and Company, 2014).

5. For a discussion of the international institutions formed during the advent of the Great Acceleration, see Will Steffen, Jacques Grinevald, Paul Crutzen, and John McNeill, "The Anthropocene: Conceptual and Historical Perspectives," *Philosophical Transactions of the Royal Society A* 369, no. 1938 (2011): 850.

6. W. Jeffrey Bolster, *The Mortal Sea: Fishing the Atlantic in the Age of Sail* (Cambridge, Mass.: Harvard University Press, 2012), 10.

7. Carmel Finley, *All the Fish in the Sea: Maximum Sustainable Yield and the Failure of Fisheries Management* (Chicago: University of Chicago Press, 2011), 9.

8. Finley, *All the Fish in the Sea*, and Jennifer E. Telesca, *Red Gold: The Managed Extinction of the Giant Bluefin Tuna* (Minneapolis: University of Minnesota Press, 2020).

9. Pope Francis, *Encyclical on Climate Change and Inequality: On Care for Our Common Home* (Brooklyn, N.Y.: Melville House, 2015). For a critique of the encyclical from the perspective of finance, see Philip Goodchild, "Creation, Sin, and Debt: A Response to the Papal Encyclical *Laudato si'*," *Environmental Humanities* 8, no. 2 (2016): 270–76.

10. Arjun Appadurai, "Introduction: Commodities and the Politics of Value," in *The Social Life of Things: Commodities in Cultural Perspective*, ed. Arjun Appadurai (New York: Cambridge University Press, 1986).

11. Nicole Shukin, *Animal Capital: Rendering Life in Biopolitical Times* (Minneapolis: University of Minnesota Press, 2009).

12. Pierre Bourdieu, *Algeria 1960*, trans. Richard Nice (New York: Cambridge University Press, 1979), 7 (emphasis added).

13. Randy Martin, "The Twin Towers of Financialization: Entanglements of Political and Cultural Economies," *The Global South* 3 (2009): 122.

14. Bruno Latour, *We Have Never Been Modern*, trans. Catherine Porter (Cambridge, Mass.: Harvard University Press, 1993), 68, and Munn, "The Cultural Anthropology of Time."

15. Although anthropologists offer various ways to interpret time, some contradictory, the broadest area of agreement is its social aspect, or the ways in which social experience defines what form time takes and what meaning or relevance it has in any given setting. See Greenhouse, *A Moment's Notice*, 23–25.

16. Nancy D. Munn, "Gowan Kula: Spatiotemporal Control and the Symbolism of Influence," in *The Kula: New Perspectives on Massim Exchange*, eds. Jerry W. Leach and Edmund Leach (New York: Cambridge University Press, 1983), 280 (emphasis in original).

17. Munn, "The Cultural Anthropology of Time," 109.

18. Greenhouse, *A Moment's Notice*, 4.

19. Greenhouse, 20–21.

20. Latour, *We Have Never Been Modern*, 68.

21. Greenhouse, *A Moment's Notice*, 22–23.

22. Greenhouse, 22.

23. Jennifer E. Telesca, "Consensus for Whom? Gaming the Market for Atlantic Bluefin Tuna through the Empire of Bureaucracy," *Cambridge Journal of Anthropology* 33, no. 1 (2015): 49–64.

24. Greenhouse, *A Moment's Notice*, 13.

25. Greenhouse, 20.

26. Widely cited in the press, this play on ICCAT's acronym first appeared in Carl Safina, "Bluefin Tuna in the West Atlantic: Negligent Management and the Making of an Endangered Species," *Conservation Biology* 7 (1993): 229–34.

27. The European Union with its twenty-seven member states constitutes one of ICCAT's Contracting Parties, or signatories to the ICCAT treaty. In 2011, at the time of the ICCAT Commission meeting in Istanbul, nearly seventy-five of the world's countries participated in the ICCAT regime from regions well beyond the Atlantic Basin. For a list of Contracting Parties see https://www.iccat .int/en/contracting.htm (accessed May 3, 2017).

28. The ICCAT Commission granted observer status to the Institute for Public Knowledge (IPK) at New York University in October 2010. The author attended ICCAT meetings as part of IPK's delegation for three years.

29. Only four of thirty-eight semi-structured interviews conducted during my research were completely on the record. Although this diplomat was not one of these interviewees—we met in passing—I do not name delegates in my work for their protection.

30. Telesca, "Consensus for Whom?" 52–54.

31. For a fascinating discussion of how ecological timescales differ on land and at sea—which fisheries management cannot accommodate—see Bolser, *The Mortal Sea,* 17–20.

32. Jennifer E. Telesca, "Accounting for Loss in Fish *Stocks*: A Word on Life as Biological Asset," *Environment and Society: Advances in Research* 8, no. 1 (2017): 144–60.

33. P. A. Larkin, "An Epitaph for the Concept of Maximum Sustainable Yield," *Transactions of the American Fisheries Society* 106, no. 1 (1977): 2.

34. See Larkin, "An Epitaph for the Concept of Maximum Sustainable Yield," and Sidney Holt, "Maximum Sustainable Yield: The Worst Idea in Fisheries Management" (2011), available online at https://breachingtheblue.com/2011/10/03 /maximum-sustainable-yield-the-worst-idea-in-fisheries-management/ (accessed January 2, 2018).

35. Jennifer E. Telesca, "Volatility," Theorizing the Contemporary, *Cultural Anthropology* website (June 27, 2018), available online at https://culanth.org /fieldsights/1464-volatility (accessed January 11, 2019).

36. Despite the intricate performance of quantitative expertise, ICCAT has only recently followed the advice of its scientific committee. See Telesca, *Red Gold.*

37. Martin, "The Twin Towers of Financialization," 111.

38. Martin, 113. The extent to which the Kobe Matrix signals a mode of economy or "moment in the genealogy of capital" known as financialization, which takes as its central instrument the derivative, is a topic for further study. See Martin, 109.

39. See the following link: https://www.iccat.int/Documents/Recs/compendiopdf -e/2011-14-e.pdf (accessed April 30, 2017).

40. Latour, *We Have Never Been Modern,* 76 (emphasis in original).

41. Martin, "The Twin Towers of Financialization," 119.

42. Bolster, *The Mortal Sea*, 7.

43. Deborah Cowan and Neil Smith, "After Geopolitics? From the Geopolitical Social to Geoeconomics," *Antipode* 41, no. 1 (2009): 22–48.

44. Telesca, *Red Gold*.

45. Greenhouse, *A Moment's Notice*, 5.

46. Greenhouse, 86.

47. The most recent report by the Food and Agriculture Organization (FAO) of the United Nations, "State of the World Fisheries and Aquaculture 2020" is available online at http://www.fao.org/state-of-fisheries-aquaculture (accessed June 30, 2020).

48. Fishing nations commonly underreport catch. See Daniel Pauly and Dirk Zeller, "Catch Reconstructions Reveal That Global Marine Fisheries Catches Are Higher than Reported and Declining," *Nature Communications* 7 (2016): 1–9.

49. Nancy D. Munn, *The Fame of Gawa: A Symbolic Study of Value Transformation in a Massim (Papua New Guinea) Society* (Durham, N.C.: Duke University Press, 1992 [1986]), 11.

50. Christophe Bonneuil and Jean-Baptiste Fressoz, *The Shock of the Anthropocene* (New York: Verso, 2017), 21.

51. Bonneuil and Fressoz, 203.

52. Greenhouse, *A Moment's Notice*, 49.

53. See Greenhouse, *A Moment's Notice*, for a discussion of resistance—and adaption—as a form of linear time.

54. Audrey Lourde, "The Master's Tools Will Never Dismantle the Master's House," in *Sister Outsider* (Berkeley, Calif.: Crossing Press, 1984), 110–13.

55. Anna Tsing, "Earth Stalked by Man," *The Cambridge Journal of Anthropology* 34, no. 1 (2016): 2–16.

56. See Martin, "The Twin Towers of Financialization," 108.

ETUDE 2
WetLand

PLATE E2.1. Mary Mattingly, Docking *Wetland* at Bartram's Garden, Philadelphia (2016). Photograph by Phil Flynn.

PLATE E2.2. Mary Mattingly, *WetLand* down Philadelphia's Schuylkill River (2016). Photograph by Phil Flynn.

CHAPTER 9

WetLand Manifesto

MARY MATTINGLY

A proposal for ascetic living in urban nature, *WetLand* was an eco-system on a boat that was simultaneously theater, home, and social sculpture. Established in 2014, *WetLand* was conceived as a public floating sculpture and artist residency on the Lower Schuylkill River in Philadelphia, Pennsylvania. An architectural folly, it was built on an old 45-foot-long Rockwell Whitcraft houseboat hull to resemble a partially submerged building, and it functioned as an experimental workspace for the Anthropocene. The wood from a public school's gymnasium floor and fifty-five-gallon drums that were used to transport syrup typified the building materials *WetLand* was constructed from.

Through a core team of six systems designers and builders, its onboard amenities included a contained living system that produced food and electricity, while it collected and filtered rainwater. Functioning like a tiny island, *WetLand* maintained onboard gardens that fed livestock, bees, and humans, while their leftover compost fed gardens. It was experienced on a scale that made it plausible for residents to directly feel their impact and for visitors to visualize a cause and effect: when a resident used a less biodegradable soap to shower, the plants in the greywater system wilted. On the water, physical connections occurred: residents responded to the felt presence of neighbors who passed by in ships and boats, and depended on the weather, which directly affected the vessel and everyone onboard.

Ecosystem Interdependence

While *WetLand* had a direct connection with the water that surrounded it, its connection with the city was equally important. Since it was not intended to be a closed ecosystem but rather function interdependently within the city, we orchestrated partnerships with organizations we valued in order to exchange for surplus goods when it launched. Artists worked with urban farms to retrieve some of their extra produce, and hosted public dinners onboard when they had their own surplus. Artists used *WetLand* as a studio from which to research and create. The residents hosted public performances, theater pieces, and workshops, along with other open-forum events. They studied the water, and helped maintain the ecosystem. They gardened, fed chickens, took canoes out into the river, and built networks of sensors to study the soil. In return, *WetLand* became a space that was constantly evolving, being added to and taken from.

In its second year, I looked to the interdisciplinary Penn Program in the Environmental Humanities (PPEH) in order to jointly redesign, rebuild, and maintain *WetLand*. It became a classroom for students led by PPEH and a team organized by Dr. Bethany Wiggin that included architect Kate Farquhar. It hosted young people who led citizen science projects and artists who designed experiential installations, learning centers, and wildlife habitats on the water.

The alliances that stewarded *WetLand* are small examples that stress the large importance of urging more people to be involved in caring for a common home, a notion that expands to stewarding water and lands so that they may continue to be safely lived with in multiple ways.

Ethos Behind Building *WetLand*

Dr. Elinor Ostrom and others taught about successful commoning by sharing examples of sustainably managed common spaces. My personal drive to increase participation in a commons came partly from growing up in an agricultural town whose inhabitants were devastated with illness from public drinking water poisoned with pesticides.

WetLand argued for valuing collective maintenance and repair: of the water, land, vessel, as well as our relationships to each other. Through narration, the layered and accumulative histories of the materials that compose *WetLand* became stories about transformation. Listening to the lives of objects, and to how they have influenced the flora, fauna, and humans they've touched, drove home a need for collective management of common spaces. Through building and living on *WetLand* and later seeing how it grew without me, I was able to understand how de-siloing allowed for robust collective imagining of leverage points that could reverse extractive systems. It also helped me understand that we need collaborative visioning to heal from damage to human and environmental health, and that a continued awareness of our fragility and interdependence will leave less room for indifference.

CHAPTER 10

Figuring *WetLand*

KATE FARQUHAR

WetLand was an experimental floating lab that generated art, study, and conversation. It was originally created from a forty-five-foot-long salvaged houseboat in 2014; then artist Mary Mattingly and collaborators added a makeshift loft bedroom and amenities, including a composting toilet, solar array, and rain-gathering cistern that redistributed water to planters and sinks. Adjacent rafts containing floating gardens supported vegetables, wetland plants, and, at times, bees and chickens. Transported from one venue to the next by tow-barges (due to an inoperable motor), it served as a sustainable showcase, event space, artist residency, and architectural folly. From 2015 to 2016, I served as the program coordinator for events that accompanied its residency with the Penn Program in Environmental Humanities (PPEH) on the Lower Schuylkill River. This is the story of its baptism, inhabitancy, and eventual surrender within the Lower Schuylkill River. With this retelling, I will focus on *WetLand's* time and legacy on the Lower Schuylkill. Like any account that involves adventuring and draws on multiple lived experiences, this record is one among many.

WetLand as Focused Spyglass

WetLand was delivered to the Lower Schuylkill, its temporary home off the bank at Bartram's Garden, by a tow barge in fall 2015. Its scant crew included Danielle Redden, Bethany Wiggin, Phil Flynn, and their guide, Bob Smizer of Smizer Boat Hauling. The group traveled slowly on waters that are commonly viewed from above while transiting between regional hubs: Philadelphia's two major arterial highways and its international airport. On the lower Schuylkill riverbanks, vast tracts of filled wetland now support refinery facilities operated by Philadelphia Energy Solutions (PES).[1] In aggregated industrial area,

PES property constitutes the largest refinery on North America's East Coast and is expected to grow. This industrial riverside landscape also presently serves up some of the worst air quality problems in the nation, causing high rates of asthma, especially in surrounding neighborhoods.[2] Though vast and visible, the huge tracts of PES land are nevertheless remote—separated from the city fabric by guarded fences. In many areas, PES's barriers prevent Philadelphians from viewing or walking to the river. A rare exception, Bartram's Garden is a popular historic landmark that offers precious access to the river and echoes the grandeur of scenic riverside parks upstream. Half a mile of planted sloping banks offset the eight and a half adjacent miles of PES's industrial shoreline, which flank the Schuylkill downstream. On the trip to install *WetLand*, the crew got as close as one can get to the flat, long, paved no-place occupied by North America's last mega-refinery on the East Coast. Guiding *WetLand*—a speculative, ecotopian floating home—to its residency required passage through a gauntlet of riprap, past Philadelphia's most exploited and inhospitable landscape.

WetLand as Stiff Negotiation

In the months leading up to *WetLand's* arrival, the Penn Program in Environmental Humanities (PPEH) engaged in talks with Philadelphia Parks and Recreation and the University of Pennsylvania to permit *WetLand* to be accessible to the public for a brief series of events. With less than twenty-four hours to spare before the first scheduled event, PPEH secured permission to hold the event series at Walnut Street dock, a small area alongside a popular river trail where a narrow gateway and ramp occasionally opened to allow access to craft floating on the Schuylkill's tidal waters.

At first, those of us working to organize events were surprised that a very small access point to Philadelphia's most visible and popular body of water could be so heavily guarded by bureaucracy. However, over the months, these negotiations would become increasingly familiar. Liability coverage for the events—including moving *WetLand* to an accessible place, hosting welcoming activities, and inviting visitors to the boat—had to be evaluated and approved by the insurers of all associated institutions. Finally, on the first day *WetLand* opened to the public, we found the gate to the hard-won ramp still locked. After we gave up calling an unresponsive Parks and Recreation number, we all teetered back and forth over the locked gate to move between *WetLand* and the trail, and host our event as planned.

WetLand as Maintenance Art

Mattingly's manual for living in *WetLand* opened with the instruction, "Welcome to *WetLand*! Make yourself at home. This is your space. Feel free to make it your own. Here are some basic instructions for taking care of *WetLand* so it takes better care of you!" The document was a written tour guide of the small living space, which included sleeping cabins on the top deck, a main event space, a small mess kitchen, hull and access hatches, shower and toilet. Instructions for managing the shower and toilet identified them as stopping places in a more expansive ecological sequence, as working organs that fed *WetLand* and its surrounding gardens:

> The shower system is based on rainwater collection, and the 40 gallon cistern is above the kitchen roof. Water collects from gutters, shoots through a screen, and into the cistern. Turn the tap on in the tub and this water should be warm by virtue of the dark color hose and tank, as well as being encased in Plexiglass. To use the shower, turn on the hose inside of the tub. Make sure that the tub's water is running to a planter bed, and that you are using a castile soap so as to not harm the plants . . . Using the toilet is also simple. Urine is diverted to a forward bottle for separate emptying. Solid waste is captured in the bowl and sent "south" into the composting tank via a manually operated trap/flap . . . When full, you can empty the tank into a land based composting toilet or put the tank lid on and take the tank home and leave it in storage. In 34 months it'll be ready to use on nonedible plants.

The tiny barge was packed with transformational tools for living, and each component, from the portable kitchen stove to the refrigerator and solar equipment, was selected with care. Despite the comprehensive user's guide, *WetLand* arrived with some unexpected quirks and demands. These included intermittent mold, bilge pumping, and several craggy spots where the salvaged wooden fascia had warped from exposure to the elements. But on the eve of its public debut on the Lower Schuylkill, my maintenance task was simple: tidy up for the opening party. Hustling to the dock after work, I marshaled cleaning supplies, a headlamp, and some book donations to *WetLand*'s library shelves, which I had pulled from my own. After boarding *WetLand*, I set to work cleaning the windows, scouring the kitchen area, sweeping floors, and staging library books while dusk settled.

WetLand's instructional manual came with a companion document called "Emergency Operations Procedure for Docked *WetLand* Barge." Emergency operations included sensible advice for tangled situations such as "Maritime Danger," "Disruptive Individuals," "Fire," and finally "Security Concerns During Pope Francis's Visit." Fortunately none of these emergencies arose during its inhabitancy of the Lower Schuylkill. However, the question that *WetLand* continued to provoke among its closest circle of stewards was how to create opportunities to engage with an artwork that was most easily described as a "sinking house."[3] Given its mounting and increasingly complex maintenance demands, the answer to this question became a moving target.[4] However, that night was still early on in *WetLand*'s residency, and my time alone cleaning and decorating was easy and satisfying. I rang the dinner bell hanging at the boat's entryway as I left.

WetLand as a Way of Knowing

In the latter half of October 2015, *WetLand* became the focus for a whirl of events: dramatic readings were performed by clusters of Penn students, the art collective We The Weeds led a cocktail-making workshop that used elixirs derived from invasive plants, Danielle Toronyi of the environmental noise group Peak Discharge collected sounds from the Schuylkill's tidal banks, local horticulturists Kylin Mettler and Robin Rick hosted a plant clinic, artist Jacob Rivkin and I made an instructional video about native seed bombs. Through all of these events, porch talk flowed between acquaintances and strangers: an easy current of chitchat, information, impressions, and advice delivered in a nonlinear, nonhierarchical fashion. Communication was carried by its setting and shaped by circles of busy hands, infused by sensations of bobbing, eavesdropping, molding, sipping, bracing, balancing, basking, and shivering. The viscera got involved.

These weeks were a time of reckoning with larger conditions of space and time. Toronyi's collection of sound was predicated on the knowledge that that October was the warmest on record, conditions that brought about a cascade of small and large local (but globally connected) consequences that could be captured sonically. Another consequence of these unusual climatic conditions, the fall of 2015 yielded an extravagant harvest of fruit, nuts, and berries—a mast year—with beneficial repercussions for wildlife. The events housed at *WetLand*—broadcasting planting tips and native seeds, testing the waters, and comparing memories of weather—all experimented with modes of perceiving, doing, and knowing.[5]

WetLand as Midnight Snack

One autumn night after *WetLand* had been moved to its hibernation spot, moored several yards off the bank of Bartram's Garden, a couple students and I headed to a planned sleepover aboard the craft. We packed cider and donuts, sleeping bags, and headlamps onto a little raft to see us through our adventure in comfort. As we approached *WetLand*, we heard clattering and glimpsed flashes of light. To our surprise, we encountered solar engineer Rand Weeks aboard. Rand had designed the solar integration with *WetLand's* electrical system, and when we met him that night he was entwined among flashlights and wires in the main cabin space. A practitioner inclined toward marathon working sessions in cloistral seclusion, Rand humored us by introducing the world-class battery that fed *WetLand's* electrical system. We wiled away the hours sharing stories and learning about other adventure craft that Rand had outfitted. As we reached the end of our provisions, we heard a shout from the dark water outside—it was a solo sailing hobbyist who was also exploring the Lower Schuylkill on his boat that night. He asked if he could moor alongside us and visit, so we poured him a cup of cider and he joined our late-night chat.

WetLand as Convener

In the spring of 2016, PPEH convened a meeting to gauge interdisciplinary interest in the Lower Schuylkill, review ongoing scholarship of the river's social and environmental histories, and build energy around citizen-science initiatives. Plans were launched to generate an "open tour" mobile app to better disseminate information about the river and connect with nonacademic publics. *WetLand* was a centerpiece of these discussions, and Mattingly expressed her hope that *WetLand* would continue to offer a space for lateral—instead of hierarchical—exchanges of knowledge.

This first convening grew into a collaboration of engaged stakeholders—the Lower Schuylkill River Corps—and an ongoing presentation series called the River Research Seminar. The lateral and open movement of knowledge has remained a core value throughout those conversations.

WetLand as Nesting Ground

Late winter 2017 found *WetLand's* stewards gearing up for the assimilation of Ecotopian Tools into its small environmental niche.[6] Ecotopian Tools was conceived as a public event series and design competition. In the spring, seven winning projects were selected out of a pool of proposals to engage the river banks in ecotopian inquiry. The winning artists facilitated community events, created environmental prototypes, and designed thoughtful tools for local use. Winning toolmakers included Cecily Anderson, Joanne Douglas, Carolyn Hesse, Gabriel Kaprielian, Mandy Katz, Jacob Rivkin, and Eric Blasco. As a kickoff to the toolmaking season, the artists and facilitators came together to renovate the garden docks flanking *WetLand*. Our group got to work with potting soil, lumber, marine-grade screws, zip ties, native plants, and landscape fabric. Hardy irises and sedums that survived from the first installation were carefully replanted, lumber was cut into bespoke jigsaw-like arrangements to fit snugly around not-quite-square docks, and new planting and thorough watering completed the job. In their first week back on the water, the gardens were visited by nesting geese who demolished the perennial grass seedlings we planted. We hopeful gardeners shared photos of eggs and damaged planting, conjecturing that the geese, like so many wildlife seeking intertidal habitat on their migratory routes, were hungry for native plants to eat and desperate for a calm, soft place to raise their young by the river. One day the carefully guarded nest was found emptied of its brood—they had slipped away from us unnoticed.

WetLand as Floating Culture

The Ecotopian Tool projects activated a curious culture around *WetLand* and the Schuylkill banks nearby. Enriched by an energetic audience during Bartram's free boating days, the tools complemented routine, warm-weather activities along the water such as fishing, boating, and various directed and undirected modes of engagement with the river. The Ecotopian Tools for *WetLand* design competition began as a call for proposals and juried selection process that yielded seven winners. Conceived as speculative "prototypes," the winning creators made tools that spoke to the local realities of climate change in Philadelphia. Cecily Anderson developed a comprehensive map of the Lower Schuylkill after collecting input from boaters and other community members about what information they most valued. Joanne Douglas used fibers and dyes to register environmental information along the river bank, and offered tutorials to curious visitors. Carolyn Hesse built a floating, mirror-like installation, called SUSPEND to animate the river's surface at dusk. The floating sculpture was placed near the ruin of a historic train bridge, which is scheduled to be replaced in coming years with a new pedestrian crossing. After leading a community "plant-walk" to identify key wetland plants, Mandy Katz began work on a comprehensive field guide to riverside plants at Bartram's Garden. Gabriel Kaprielian, with collaborators Kya Kerner and William Bourke, created floating modules that hosted native intertidal plants and bore flags featuring silhouettes of native river animals. Jacob Rivkin and Eric Blasco built a "bio-pool"—a passive floating form that used planted bio-char panels—to cleanse the river water passing by. A series of walks and workshops featuring each tool generated new conversations between friends and strangers, and gasp-worthy moments of splendor.

One such instant was when Rivkin and Blasco's volunteers shoved the bio-pool's land-locked framework forward to crash (and float!) onto the Schuylkill's surface. Similarly breathtaking was the sight of the setting sun glinting off the intricate panels of SUSPEND, and noticing the delight people experienced when their shadows altered photochromatic fibers Douglas prepared. These happenings were the ethereal framework of a floating society, punctuated by convivial conversation and shimmering outbursts of human imagination.

WetLand as Folly

The unusual sight of *WetLand* became routine throughout spring 2017. Our neighborhood "sinking house" was moored, so that from the shore one usually saw its side, partially screening the floating gardens and ecotopian tools assembled beyond. The little fleet rose and sank twice daily with the tide, constantly swaying and pivoting on independent axes. Signs of wear-and-tear from *WetLand's* seafaring years accumulated and became increasingly visible to passersby: the railings seemed to need re-attachment and at least one window was broken. One afternoon as I watched its slow pivoting movement in the water, Danielle Redden (who led public programs at Bartrams' Gardens Community Dock) said to me, "think of how far it has traveled—imagine if you towed your dresser thousands of miles across open water."

As I reflect on short and long conversations about *WetLand*—sometimes as short as, "What IS that?"—I picture *WetLand* as a folly in the landscape. In the eighteenth century, during the Romantic craze in Western Europe, elite garden landscapes were populated with small ornamental structures called fabriques or follies. Not intended for practical use or as living quarters, they served many symbolic purposes: as theatrical scenery to role-playing games, as objects of meditation to feed "solitary contemplation and quiet friendship," and finally as fodder for reverie.[7] Redden and I exchanged memories of unusual interactions we shared throughout *WetLand's* residency, and of the kinds of contact people sought with it. She recalled countless interactions that went something like this: Q: "What is that?!" A: "It's a boat and an art project that is visiting the Lower Schuylkill." Q: "Does anyone live there?" A: "No." Q: "Can I live there?" For all its functional quirks, *WetLand* transmuted quite fully into the folly role. Participants in the free boating program at Bartram's saw the boat from two differently captivating angles. It appeared as a bobbing novelty when seen while strolling along the Schuylkill River Trail.

But from a rower's vantage, it made an iconic protrusion from the water. *WetLand* also functioned as a party centerpiece for two festive Penn Program in Environmental Humanities' academic celebrations. Youth seemed to engage with *WetLand* as target practice. In fact, *WetLand*'s moor lines were cut by unknown parties an astonishing number of times. The most notable incident occurred on Christmas 2016, when Redden received a call from Philadelphia police that *WetLand* was drifting along the Lower Schuylkill. Many steps were required to cut *WetLand* loose, so we were amazed that it happened at all, let alone several times. To quote Redden, "People like boats, and it's not that common for people to see or contact boats here so it's special . . . It seems like people just wanted to set *WetLand* free."

WetLand Is Not *Spiral Jetty*

On the morning of August 6, 2017, I received a text of an image of *WetLand* with just the tip of its roof peeking above water. Soon after, a long, hard rainstorm blew in. Against the soundtrack of the beating rain, Mary Mattingly and our cluster of local stewards discussed what to do. The following week, after exploring several creative salvage options, we scheduled *WetLand*'s final tow to a conventional salvage yard. The final voyage was preceded by a spectacular twenty-four-hour procedure in which expert divers installed airbags under the craft and inflated them to elevate *WetLand*, reducing drag and allowing it to drain, lessening the load to tow. I convinced a friend to pay a visit during those final hours and photograph the flotation procedure, and scheduled a subsequent visit to see *WetLand* off. Unfortunately, the choreography of events that included the flotation procedure, tow, and scrapyard reception found their own momentum; those of us who arrived to witness *WetLand*'s final voyage found it had already departed. Only the small fleet of offspring remained: the replanted floating gardens, Rivkin and Blasco's bio-pool, and an array of rowboats used on Bartrams' free boating days.

More than any other memory of *WetLand*, its sinking stands out to me as a moment so stark it defies mention. I am tempted to point away from myself, toward a summary of other accounts, such as those by its creator, from news articles, and curators. A sinking house / An ode to Katrina / A place for the assimilation and surrender of resources. This cataclysm was written into its fate—what else could I possibly add?

From a historical perspective, the sinking of *WetLand* raises the query of whether this artwork will join the pantheon of projects that confront ruination and riff on the second law of thermodynamics. Works come to mind such as Gordon Matta-Clark's "building cuts" and Robert Smithson's *Partially Buried Woodshed*. Furthermore, the seduction of ruination pervades contemporary popular culture: Burning Man, #RuinPorn, and apocalyptic stories such as Alan Weisman's bestselling thought-experiment-as-book *The World Without Us*.

Robert Smithson's *Spiral Jetty* offers a useful counterpoint—soon after it was built (during drought), it became submerged. It later re-emerged in the 1990s, staying more reliably above water. This dual identity, as a sharply crystallized or obscured, dissolved work, has been one of the most iconic characteristics of its geographically remote slip. Since the 2000s, water levels have progressively lowered as the western United States has entered a period of purported "megadrought," and lake levels have dropped due to water demands upstream. The progress of expected drought conditions will likely leave the artwork marooned in dry sand, which will eventually change its aesthetic qualities. Though the site's red ooze—produced by halophilic organisms—attracted Smithson, these too will eventually fade out of the salt crystals left on dry land.[8]

WetLand has been spared the slow bake that Smithson's Jetty endures, but I want to tell you how it felt to witness *WetLand* go, rather than summarize its potential place in the record. A cozy, collaborative craft that speaks to ecotopian futures (in my case, one of my books that sank in its library was a collection of Futurist manifestoes), many of us put time and treasure into *WetLand*. At *Spiral Jetty*, the effects of climate change and human demands are made visible in the slow transition of a sublime vista, whereas *WetLand* is known as a homeplace. The day it sank showed us in an instant how climate disaster reaches into and disrupts daily living. The event was witnessed by an audience of two. Carolyn Hesse, who had built SUSPEND for the Ecotopian Toolkit series, was out kayaking with a friend. She told us later it plummeted to the silty bottom in a matter of seconds.

WetLand as Ouroboros

By the mid-nineteenth century, North American cities were actively adopting England's sanitation model—in which civic infrastructure delivered drinking water to households and controlled domestic wastewater disposal. The progenitor of the sanitary idea was Edwin Chadwick, a barrister-turned-sanitation-administrator who proposed idealized processes for condensing resource management into a single loop.[9] Philadelphia's once-revolutionary hydrologic infrastructure followed this model, but as it has aged, one aspect has caused ongoing pollution problems in the twenty-first century. Although stormwater is supposed to be channeled directly into the Schuylkill and Delaware Rivers, those rain events that precipitate more than two inches of rainfall cause the antiquated (undersized) wastewater sewer system to overflow, in turn combining with Philadelphia's sewage and pouring into its receiving water bodies. Through an initiative called Green City/Clean Waters, which coordinates stormwater remediation efforts between the City of Philadelphia, the Philadelphia Water Department, and the Environmental Protection Agency, new greening projects are springing up throughout Philadelphia to infiltrate rainwater where it falls. The goal is to reduce dependence on the sewer system as a means of funneling stormwater, thereby keeping sewage waste on course to designated processing plants and out of rivers.

The engine behind Philadelphia's present green infrastructural improvement and long cleanup effort is the 1972 Clean Water Act, which regulates pollution discharge into bodies of water, and asserts high standards for "fishable, swimmable" use by citizens. The Schuylkill River is intercepted by numerous outflows from the sewer system as it cleaves through Philadelphia, and as a result the public boating program at Bartram's Garden pauses after rain events to mitigate contact with sewage effluent. On the tidal lower Schuylkill, the collected drops from our watershed and the pollution they incorporate merge with the brackish waters of the Delaware estuary. At this crucial ecological threshold for migrating animals and plant communities reliant on intertidal conditions, *WetLand* was poised for a time. Initially it showcased the tools for sustainable living, such as redirecting human waste into more expansive ecological partnerships. But at the close of its residency it unexpectedly joined the conventional waste cycle in Philadelphia's scrap yard, stripped for parts and material commodities. She was a ponderous charge, a chorus of aqueous reveries and an imperfect demonstration of resource management: but in the end—like the mythical beast Ouroboros—the creature's mouth found its tail and kept spinning.

NOTES

1. As of publication, the PES refinery in south Philadelphia has closed permanently, following explosions in 2019. Industrial usage of the land is expected to continue.

2. Asthma and Allergy Foundation of America, *Asthma Capitals 2015* (Landover, Md.: AAFA, 2015), and Pennsylvania Department of Health, *2015 Pennsylvania Asthma Prevalence Report* (Harrisburg, Penn.: PADH, 2015).

3. Mary Mattingly said, "We're in this really dystopian time," she said, citing the devastation of Hurricanes Katrina and Sandy. "To me, the sinking house can be iconic. It's really just what's been happening all over the East Coast recently." S. Melamed, "*WetLand,* a Floating—and Sinking—Artwork on the River," August 14, 2014, http://www.philly.com.

4. "Avant-garde art, which claims utter development, is infected by strains of maintenance ideas, maintenance activities, and maintenance materials. Conceptual & Process art, especially, claim pure development and change, yet employ almost purely maintenance processes." Mierle Laderman Ukeles, *Maintenance Art Manifesto 1969*, Feldman Gallery website, https://feldmangallery.com /exhibition/manifesto-for-maintenance-art-1969 (accessed October 1, 2017).

5. "The rise of reductionist science was linked with the commercialization of science, and resulted in the domination of women and non-Western peoples. Their diverse knowledge systems were not treated as legitimate ways of knowing. With commercialization as the objective, reductionism became the criterion of scientific validity. Nonreductionist and ecological ways of knowing, and nonreductionist and ecological systems of knowledge, were pushed out and marginalized." Vandana Shiva, "Can Life Be Made? Can Life Be Owned?" in *The Vandana Shiva Reader* (Lexington: University Press of Kentucky, 2014), 143.

6. "Living systems are units of interactions; they exist in an ambience. From a purely biological point of view they cannot be understood independently of that part of the ambience with which they interact: the niche; nor can the niche be defined independently of the living system that specifies it." H. R. Maturana, *Biology of Cognition* (Ft. Belvoir, VA: Defense Technical Information Center, 1970), 9.

7. Elizabeth B. Rogers, *Landscape Design: A History of Cities, Parks, and Gardens* (New York: Harry N. Abrams, 2001), 261–63.

8. A. Wang, "As the Great Salt Lake Dries Up, 'Spiral Jetty' May Be Marooned," February 7, 2017, https://hyperallergic.com

9. "For Chadwick, the appropriate response for dealing with unhealthy conditions was to be found in improved public works, including waterworks, sewers, paved streets, and ventilated buildings. He proposed a hydraulic (or arterial-venous) system that would bring potable water into homes equipped with water closets, and then would carry effluent out to public sewer lines, ultimately to be deposited as 'liquid manures' onto neighboring agricultural fields . . . With the addition of the fertilization phase, Chadwick noted, 'we complete the circle, and realize the Egyptian type of eternity by bringing as it were the serpent's tail into the serpent's mouth.'" Martin Melosi, *The Sanitary City* (Pittsburgh: University of Pittsburgh Press, 2000), 31–32.

PART III
REPETITIONS AND VARIATIONS

CHAPTER 11

Vanishing Sounds
Thoreau and the Sixth Extinction
WAI CHEE DIMOCK

My essay is a meditation on Thoreau in the context of climate change and the loss of biodiversity, in the nineteenth century no less than the twenty-first. The time scales that I'll be exploring require a two-fold chronology, a contrapuntal structure with two temporal homes, shuttling back and forth between past and present the better to track the phenomenal fields extending between them. Evolution, extinction, and reparation are most meaningfully investigated, I would argue, within this twofold temporal structure. In that spirit I begin not with Thoreau himself but at some distance from him, with a recent work on the loss of biodiversity as a sonic phenomenon, Bernie Krause's *The Great Animal Orchestra* (2012).[1]

Krause is something of a cult figure to music fans: the last guitarist recruited by the Weavers to replace Pete Seeger, he teamed up a bit later with Paul Beaver to form the legendary synthesizer team, Beaver and Krause, providing electronic music for films such as *Rosemary's Baby* and *Apocalypse Now*. For the past forty years, though, his work has been primarily in bio-acoustics, focusing especially on the sound ecology of endangered habitats. *Wild Sanctuary*, his natural soundscape collection, now has more than four thousand hours of recordings of more than fifteen thousand species.[2]

Krause tells us that animals consistently outperform us when it comes to sound: they both hear and vocalize better than we do and can also do more with sound than we can. One example he gives is the sound camouflage perfected by the spadefoot toad. This amphibian species, like many animals in the wild, does not vocalize separately, but does so as a group, "a synchronous chorus assuring a seamless protective acoustic texture."[3] Through this sound aggregation, they prevent predators such as foxes, coyotes, and owls from pouncing on one particular victim, since no single individual stands out.

Unfortunately, the complexity of this camouflage is such that any human interference, any artificially generated noise that falls outside the usual sound spectrum within this particular environment, is likely to disrupt it and undermine its working. The spadefoot toad is a case in point. Krause starts with their marvelous sound engineering, but by the time he is done it is no longer a happy story. When a military jet flew "low over the terrain nearly four miles west of the site," the sound camouflage was thrown off kilter. It took the toads between thirty to forty-five minutes to rebuild it. Krause reports: "My wife and I watched from our nearby campsite as a pair of coyotes and a great horned owl swept in to pick off a few toads during their attempts to reestablish vocal synchronicity."[4]

The death of a few spadefoot toads is probably no major disaster, but the larger narrative that comes out of *The Great Animal Orchestra* is disturbing in more ways than one, with an intimated ending that probably none of us would want to hear in full. Something much larger, more systemic, and more destructive than military jets is preying on these sound ecologies, upsetting their delicate balance and making them less and less able to function as they used to. Almost half of the habitats in which Krause made his recordings have now been seriously compromised or destroyed. His audio archives are all that is left of those once sound-rich environments.

Bernie Krause is writing at a point in time when environmental degradation has led to a mind-boggling list of plant and animal species that have gone extinct. Turning from this book to *Walden* is not only a journey back in time; it is also an act of reckoning, a survey of the damage done, and, I hope, the beginning of a strategy for repairs. To devise a way forward, our increasingly fragile environment needs to be seen against one that was still relatively robust, beginning to be adversely affected, but also holding the hope of a different developmental pathway, pointing to what needs to be done to ensure collective survival into the future.

Frogs Then and Now

There is no better example of these cross-currents than the boisterous frog chorus in chapter 4 of *Walden*:

In the mean while all the shore rang with the trump of bullfrogs, the sturdy spirits of ancient wine-bibbers and wassailers, still unrepentant, trying to sing a catch in their Stygian lake,—if the Walden nymphs will pardon the comparison, for though there are almost no weeds, there are frogs there,—who would fain keep up the hilarious rules of their old festive tables, though their voices have waxed hoarse and solemnly grave, mocking at mirth, and the wine has lost its flavor, and become only liquor to distend their paunches. . . . The most aldermanic . . . quaffs a deep draught of the once scorned water, and passes round the cup with the ejaculation *tr-r-r-oonk, tr-r-r-oonk, tr-r-r-oonk!* And straightway comes over the water from some distant cove the same password repeated.[5]

"Repeated": this is the keyword here, perhaps the single most important word in this nineteenth-century report on the natural environment. Nothing spectacular, just the sense that there will be more, that whatever is happening now will happen again, a dilation of time that makes the future an endless iteration of the present. All of this is suggested by the croaking of the bullfrog, so natural to that particular habitat and so eternal in its recurrence that it is unimaginable there would ever be a time when that sound would not be there. The future of the planet is guaranteed, and we, along with the frogs, could rest in that security.

It is a luxury, of course, to feel that way. *Walden* is very much a nineteenth-century text in this sense; there is no better measure of our distance from that world than the evaporation of that sense of security. Among the escalating changes to the environment that began with the Industrial Revolution, the loss of a bountiful and habitable future must rank near the top. The now-familiar term "Anthropocene," coined in the 1980s by ecologist Eugene F. Stoermer and atmospheric chemist Paul Crutzen, names human behavior as the chief cause for the drastically altered conditions for life on the planet, so abrupt and unprecedented as to constitute a new geological epoch. A "sixth extinction" seems well underway, the elimination of a "significant proportion of the world's biota in a geologically insignificant amount of time."[6] Such massive die-offs have happened only five times in the 3.6 billion-year history of life on the planet. Each time, it took millions of years for life to recover, starting from scratch with single-celled organisms such as bacteria and protozoans.[7] The sixth extinction—if that is indeed what we are headed for—promises to be even more cataclysmic than the previous five. The work of just one species, it has already resulted in 140,000 species disappearing each year, while half the life-forms on earth are slated for extinction by 2100, according to E. O. Wilson, writing in 2002.[8] In the ten years that followed, Wilson's predictions have been more than borne out. Elizabeth Kolbert now reports that "it is estimated that one-third of all reef-building corals, a third of all fresh-water mollusks, a third of sharks and rays, a quarter of all mammals, a fifth of all reptiles, and a sixth of all birds are headed toward oblivion."[9]

Frogs turn out to have been the first to sound the alarm. Since the mid-1970s, herpetologists from all over the world began to hear, not the loud croaking of frogs but an eerie silence, a deafening absence of sound. Researchers from North America, the United Kingdom, Australia, and New Zealand started comparing notes, puzzled by the fact that, in many of these cases, the disappearances were taking place apparently without encroaching human presence. There was no suburban development or highways with life-threatening traffic. The frogs seem to be dying out on their own.

What makes these extinctions especially worrisome is that the frog is one of the oldest species on earth. Its history is one hundred times longer than human history. This long duration suggests that the evolutionary history of amphibians is intertwined with the evolutionary history of the planet at every stage. Over their life cycles, they turn from tadpoles to frogs, moving from water to land and changing from plant eater to insect eater, so they have something to tell us about almost every kind of habitat. And because their skin is permeable, they are the first to register any environmental degradation, and to do so across the widest range of variables. It is for this reason that they are the proverbial canary in the coal mine: their well-being is also a measure of the well-being of the planet as a whole.

On December 13, 1992, "The Silence of the Frogs" appeared on the cover of the *New York Times Magazine*, accompanied by a nine-page article by Emily Yoffe, documenting the extinction or near-extinction of many amphibian species. Ten years later, an article appeared in *Proceedings of the National Academy of Sciences* with this title: "Are We in the Midst of the Sixth Mass Extinction? A View from the World of Amphibians."[10] The authors—David Wake and Vance T. Vandenburg—noted that, even though mass extinctions are unthinkably rare, based on the collective observations of herpetologists around the world, they would have to conclude that the unthinkable was indeed upon us. In 2009, PBS revisited the issue in a special program of *Nature* titled "The Thin Green Line," showing that already one-third of all amphibian species has now vanished.

In 2012, the *New York Times* reported that a killer had been identified: a fungus named chytrid, capable of wiping out entire frog populations in a matter of months.[11] Ironically, the principal carrier for this pathogen appears to be Thoreau's bullfrog, itself resistant to the fungus, and a popular food import, much in demand in the Chinatowns of New York, San Francisco, and Los Angeles. Human dietary habits, it seems, are responsible for the dissemination of this infectious agent, triggering mass extinctions as an unforeseen side effect.[12]

Such a future—in which side effects far exceed what initially triggered them—is not one that Thoreau could have imagined. And yet, there *is* something odd about his portrait of the frog, an oddly archaic diction that suggests this species could be a relic, both in the sense of having a venerable past and in the sense of being no longer in sync with the present. The frogs are "wine-bibbers and wassailers," "quaff[ing] a deep draught" from the pond, accompanied by nymphs and the river Styx. These overdone classical allusions, rather than drawing us into the worlds of Homer and Ovid, keep us instead at arm's length, highlighting our own modernity and separation from the ancient world. It is in such moments, when time becomes segmented and disjointed, that the eternal present of *Walden* ceases to be eternal and becomes more like a finite endgame. Past and future are no longer one, here. The terminal point is no longer a faithful replica of the starting point. It is worth noting in this context that, though still loudly croaking, the frogs are in fact no longer what they used to be. Try as they might to "keep up the hilarious rules of their old festive tables," the wine has "lost its flavor," and their own voices have grown "hoarse and solemnly grave." Theirs is a story of spiraling attrition, in which losses will be incurred that cannot be recovered, and points of no return will be reached from which there is no going back.

Animal Fables

And, if Homer and Ovid are unrecoverable for Thoreau, this is even more so in his relation to the ancient authors of animal fables, a genre now almost exclusively associated with Aesop (620–560 BCE), but for an avid reader like Thoreau, also beckoning to other, equally venerable traditions. These stories seemed to belong to the mythic past of humankind, passed down from times immemorial and flourishing in different languages and cultures. As Laura Gibbs points out, "The animal characters of Aesop's fables bear a sometimes uncanny resemblance to those in the ancient folktales of India collected both in the Hindu storybook called the *Panchatantra* (which later gave rise to the collection entitled *Kalila wa Dimnah*, a book which served as a source for many of the didactic animal stories in the Islamic mystical poet Rumi) and also in the tales of the Buddha's former births, called *jatakas*."[13] Aesop's fables were first translated into English and published by William Caxton in 1484. The Sanskrit stories were translated into English in 1775, as *Fables of Pilpay*. In an undated entry in his commonplace book (what he kept before he started his journals), Thoreau pays tribute to both these Greek and Hindu predecessors, but then proceeded to tell an animal story of his own, pulling away from both:

> Yesterday I skated after a fox over the ice. Occasionally he sat on his haunches and barked at me like a young wolf . . . All brutes seem to have a genius for mystery, an Oriental aptitude for symbols and the language of signs; and this is the origin of Pilpay and Aesop. . . . While I skated directly after him, he cantered at the top of his speed; but when I stood still, though his fear was not abated, some strange but inflexible law of his nature caused him to stop also, and sit again on his haunches. While I still stood motionless, he would go slowly a rod to one side, then sit and bark, then a rod to the other side, and sit and bark again, but did not retreat, as if spellbound.[14]

Pilpay and Aesop are mentioned by name, but there is in fact very little resemblance between this particular encounter with the fox and their much more conventional and easily recognizable forms of the fable. Pilpay's and Aesop's animals typically talk, and typically do so inside a frame story, coming with morals that are clearly stated. Thoreau's fox does not. Rather than fitting comfortably into a frame story, this animal is out there running wild, moving according to a logic of his own, one that makes no ready sense to humans. He is not the bearer of anything edifying, for he is himself a sealed book, an unyielding mystery. Yet the disturbance that he is producing in the auditory field is such as to make this sealed book an emphatic if incomprehensible warning. With an intelligence unfathomable to humans, this fox seems to come from a sonic universe with a grammar of its own, a "language of signs" older than civilization and older than human language itself.[15]

Sound is crucial. For even though the fox is not saying anything intelligible to humans, the auditory field here is in fact more electrifying than it would have had he been capable of speech. For Thoreau seems to go out of his way to create a sonic anomaly: this fox does not sound like a fox at all; his bark is like that of a young wolf. And he barks only when he is sitting on his haunches, while he is playing out an extended lockstep sequence with Thoreau himself. The man and the fox move strangely in tandem, two halves of the same ritual of speeding, stopping, and starting again, a dance of pursuit and flight, making humans and nonhumans part of the same rhythmic fabric.

And yet, this rhythmic fabric notwithstanding, the man and the fox are in fact not one, separated not only by the steadily maintained physical distance between them but also by a gulf still more intractable. "The fox belongs to a different order of things from that which reigns in the village," Thoreau says, an order of things that is "in few senses contemporary with" socialized and domesticated humans.[16] All that Thoreau can say about a creature so alien is that his bark is more wolf-like than fox-like, a strange sonic misalignment that seems a metaphor for just how little the fox is attuned to us, thwarting any attempt by Thoreau's ears to classify him, just as he runs counter to the rationality of human settlements in general.

Even though the fox was not hunted to extinction, it had nonetheless less in common with humans than with the wolf, a creature systematically exterminated in New England. As Christopher Benfey points out, "Among the first laws instituted by the Puritan settlers of the Massachusetts Bay Colony in 1630 was a bounty on wolves, which Roger Williams, who fled the colony for its religious intolerance, referred to as 'a fierce, bloodsucking persecutor.'"[17] According to the Massachusetts Division of Fisheries and Wildlife, the gray wolf has been extinct in the state since about 1840.[18] Sounding eerily like that silenced species, the fox points to a world no longer with us, one we have destroyed.

This is not the only occasion where a lost world is hinted at, through a mistaken sonic identity, a disorientation of the senses. Thoreau's celebrated encounter with the loon, in the "Brute Neighbors" chapter of *Walden,* features another instance. While Thoreau is single-mindedly pursuing this bird, his soundscape is once again strangely warped, haunted by what ought not to have been there. First of all, there is a curious expansion of the sensorium, once again ushering in the same land animal not otherwise to be found on Walden Pond:

> His usual note was this demoniac laughter, yet somewhat like that of a water-fowl: but occasionally, when he had balked me most successfully and come up a long way off, he uttered a long-drawn unearthly howl, probably more like that of a wolf than any bird; as when a beast puts his muzzle to the ground and deliberately howls. . . . At length having come up fifty rods off, he uttered one of those prolonged howls, as if calling on the god of loons to aid him, and immediately there came a wind from the east and rippled the surface, and filled the whole air with misty rain, and I was impressed as if it were the prayer of the loon answered, and his god was angry with me; and so I left him disappearing far away on the tumultuous surface.

The wolf is here, on Walden Pond, strictly because of an auditory conceit, a fancied likeness suggested by the ear. That likeness might not have occurred to everyone, but Thoreau insists on it. And if in the earlier episode it had already produced a break and a tear in the auditory fabric, the break and the tear here are still more pronounced. The wolf-like sound, a "long-drawn unearthly howl," let out at just the moment when the loon has "balked [Thoreau] most successfully," could in fact be quite unnerving. Even though it is meant to be joyous, an undertone of defiance, of unresolved hostility, and perhaps even of remembered pain, seems to lurk just below the surface. It reminds us that the loon too is not unlike the wolf, with peaceful coexistence between humans and nonhumans equally unlikely here.

Indeed, by the end of the nineteenth century, the loon would become locally extinct in eastern Massachusetts. Not till 1975 would a pair of loons be sighted again, nesting at Quabbin Reservoir. Today loons are listed on the Massachusetts Endangered Species Act list as a Species of Special Concern.[19] With remarkable prescience Thoreau seems to have anticipated all of this. Through the haunting of the ear, he injects an edginess, an intimation of harm, into an otherwise idyllic setting. In *Walden*, though, this intimation of harm is both deliberately staged, and, just as deliberately, allowed to subside. In this world—still relatively benign—miraculous deliverances do occur, crises do get averted. And the intervening force is literally a *deus ex machina*, in the shape of the god of loons, signaling to Thoreau to leave the bird alone. The patently contrived nature of the denouement is perhaps the point: this is not meant to be realistic or convincing. Any avoidance of harm that rests on this flimsy plot device is resting on a fiction that announces itself as such.

Here, then, is a story very different from those in Pilpay and Aesop. In those fables from antiquity, marked by relative harmonious continuum between humans and nonhumans, morals can be delivered, and death is edifying, not traumatic. In that clean, benign, and balanced universe, harm does occur but is fully rationalized by the concept of "desert": it teaches its lesson and comes to an end when the fatal consequences of a misdeed are embodied without fail by the culprit. If harm were to befall the loon, it is likely to take a very different form; and the attendant response is also going to be shrill and unappeased like that wolf howl, with nothing edifying to ground it, and nothing to hold it in check.

Old Testament against Darwin

How to give credence to that kind of sound? Neither Pilpay nor Aesop is of much help here. The traditional animal fable cannot tell that kind of story. But Thoreau is not without alternatives, even among ancient texts. As Stanley Cavell points out, the genre closest to his temperament might turn out to be one regularly encountered in nineteenth-century New England, namely, the elegiac lamentations of the Old Testament prophets, especially Jeremiah and Ezekiel.[20] These are voices speaking in tongues, crying out in the wilderness—and doing so because their faculty of hearing is exceptional, because they have received in full what "the Lord said unto me." Hearing and lamenting are of supreme importance in the Old Testament, affirming the primacy of a sound-based environment, and providing the language, the rhetorical structure, and above all the emotional hardihood to mourn publicly devastations that are large scale, that defy our common-sense reckoning. Here is Jeremiah: "For the mountains will I take up a weeping and wailing, and for the habitations of the wilderness a lamentation, because they are burned up, so that none can pass through them; neither can men hear the voice of the cattle; both the fowl of the heavens and the beast are fled; they are gone."[21]

Jeremiah only talks about all these species being "gone"—he says nothing about "extinction." He couldn't have—the word first appeared in the sixteenth century. But the biblical universe in which he operated was one in which that word would have meaning, one fully aware that the world was no longer what it used to be and would never again be what it used to be. This sense of a process already well underway, not only unstoppable but also nonbenign, is particularly striking if we compare it with the rhetoric of "extinction" that would later gain currency in the mid-nineteenth century. "Extinction" was a word that loomed large in *On the Origin of Species,* especially the pivotal chapter 4, "Natural Selection." But—true to Darwin's faith in a continually self-regenerating world—it was not a word that caused alarm. On the contrary, Darwin was reassured by this phenomenon, glad that there was this longstanding and always-reliable mechanism that would allow the world to update its inventory, getting rid of what is obsolete and ill-adapted. Natural selection would have been impossible without extinction, because this is the very means by which inferior species are eliminated to make way for superior ones: "The extinction of species and of whole groups of species, which has played so conspicuous a part in the history of the organic world, almost inevitably follows on the principle of natural selection, for old forms will be supplanted by new and improved forms." We shouldn't be unduly sentimental about species that die out, because it is to the benefit of all that "new and improved varieties will inevitably supplant and exterminate the older, less improved and intermediate varieties."[22]

Thoreau was an admirer of Darwin. He had read *The Voyage of the Beagle* when its New York edition appeared in 1846, taking plenty of notes. When *On the Origin of Species* came out in late 1859, Thoreau was among the first to read and comment on it. And yet, on one subject—the extinction of species—he was much less sanguine, much less convinced that it was a good thing, indeed much closer to the lamenting spirit of the Old Testament prophets. On March 23, 1856, he wrote in his journal:

> Is it not a maimed and imperfect nature that I am conversant with? As if I were to study a tribe of Indians that had lost all its warriors. . . . I listen to a concert in which so many parts are wanting . . . mutilated in many places. . . . All the great trees and beasts, fishes and fowl are gone; the streams perchance are somewhat shrunk.[23]

Endangered Humans

Already in the mid-nineteenth century, the world seemed like a damaged place, with vital parts of it gone and never to be recovered. Quite apart from the Old Testament prophets, where else did Thoreau get this idea: that there are some losses that are not compensated for and not folded into a story of progress, of creation through destruction? The two references here to Native Americans—that the world that was once fully there for them is no longer there for us, and that these depletions in nature are related to depletions among these Native tribes—seem to me especially significant. After all, the fate of these Indigenous populations was plain for all to see. Survival was not something they could count on, and neither could they count on the survival of the habitat in which they had been naturalized, in which they could flourish.

That much was clear. The question, though, was how to interpret this fact. There seemed to be two possibilities here. One was to see these Native Americans as stand-alone casualties: they were dying out for a reason peculiar to themselves. They were an exception, an anomaly. The very fact that they were indigenous here, naturalized a long time ago, meant that they had to be on their way out. They were misfits now in this environment precisely because they once had a self-evident relation to it. Thoreau did occasionally embrace this line of thinking. *The Maine Woods,* in fact, opens with a strangely cavalier remark about the "extinction" of Native Americans, taking it for granted: "The ferry here took us past the Indian Island. As I left the shore, I observed a short, shabby, washerwoman-looking Indian—they commonly have the woebegone look of the girl that cried over spilt milk . . . This picture will do to put before us the Indian's history, that is, the history of his extinction."[24]

The word "extinction" certainly stands out here, though in fact it was not a word typically used by Thoreau to describe the fate of humans. His preferred term was "extermination," speaking to the outcome of only one specific policy rather than the irreversible, across-the-board collapse of an entire population.[25] Still, as evidenced by its appearance here, "extinction" was a word that crossed species lines with relative ease in the nineteenth century. It is helpful to turn briefly from Thoreau to Washington Irving and Herman Melville to get a broader sense of how that word was used, what other terms approximated it or substituted for it, and how Native Americans fared in their midst.

Of the three authors, Irving was by far the most upfront about who was responsible for the looming catastrophe. White society, he wrote in *The Sketch Book,* "has advanced upon [Indigenous tribes] like one of those withering airs that will sometimes breed desolation over a whole region of fertility."[26] He was especially struck by one particular episode, the burning down of the Pequot's wigwams and the "indiscriminate butchery" that ensued. Entering the swamps where the warriors had taken refuge with women and children, the troops "discharged their pieces, laden with ten or twelve pistol bullets at a time, putting the muzzles of the pieces under the boughs, within a few yards" of their targets, so that all were "despatched and ended in the course of an hour."[27] New World history, for Irving, was very much a history of mass extinction, accomplished not by the benign hand of natural selection but by the trigger-happy hand of war. It was deplorable, but it was also *fait accompli.* Proleptic elegy was all Irving could offer at this point: "They will vanish like a vapor from the face of the earth; their very history will be lost in forgetfulness."[28]

He ended with the lamentation of an old warrior: "our hatchets are broken, our bows are snapped, our fires are nearly extinguished: a little longer and the white man will cease to persecute us—for we shall cease to exist!"[29]

Like Irving, Melville was fatalistic about the Pequots, speaking, in his case, with seeming callousness and finality about this doomed tribe, as in this offhand remark reminding us of the provenance of Ahab's oddly named ship: "the Pequod, you will no doubt remember, was the name of a celebrated tribe of Massachusetts Indians, now as extinct as the Medes."[30] Yet—and this suggests that what we are witnessing might be less a settled outcome than a pattern of oscillation and irresolution—it is also the case that, in the rest of *Moby-Dick*, beginning with the prophecy of "old squaw, Tistig, at Gay-Head,"[31] those Massachusetts Indians who are supposedly extinct are never absent at any point, in person or in spirit. In chapter 61, the first time a whale is killed, the "slanting sun playing upon the crimson pond in the sea, sent back its reflection into every face, so that they all glowed to each other like red men."[32] *Like red men*—the relation between whites and Indians is here no longer one of contrast, but one of resemblance: Stubb and his crew in the whale boat are doing what seafaring Native Americans have done since times immemorial. Could it be that the fate of these two are also intertwined? It is perhaps not accidental that at the end of the novel, as the concentric circles of the vortex carry "the smallest chip of the Pequod out of the sight," the disappearance of that doomed ship should come with this final image of Ahab's flag undulating "with ironical coincidings" over the "sunken head of the Indian at the main-mast," while

a red arm and a hammer hovered backwardly uplifted in the open air, in the act of nailing the flag faster and yet faster to the subsiding spar. A sky-hawk that tauntingly had followed the maintruck downwards from its natural home among the stars, pecking at the flag, and incommoding Tashtego there; this bird now chanced to intercept its broad fluttering wing between the hammer and the wood; and simultaneously feeling the ethereal thrill, the submerged savage beneath, in his death-gasp, kept his hammer frozen there; and so the bird of heaven, with archangelic shrieks, and his imperial beak thrust upwards, and his whole captive form folded in the flag of Ahab, went down with his ship.[33]

Contrary to the earlier, seemingly casual reference to the "extinction" of the Pequots, the story Melville tells is not the isolated demise of one Indigenous tribe, but an across-the-board catastrophe afflicting everyone, with the flag and the imperial eagle and the red arm and hammer all going down as one. Here then is proleptic elegy pushed to its outer limits: austere, implacable, allowing for no exit and no exception. Native Americans, far from being an anomaly, are here an advance warning to the rest of us, giving us a glimpse of the future in their already materialized "sunken" state.

Thoreau, equally mindful of that possibility, focused nonetheless not on the sublime drama of the imagined end but on humbler tasks of preservation, perhaps even reparation. Starting in 1847, during his stay at Walden Pond, Thoreau systematically compiled notes on Indigenous peoples both in North America and elsewhere, eventually ending up with eleven volumes, containing "2,800 handwritten pages or over 500,000 words." These "Indian Books," written "in English, French, Italian, Latin, and occasionally Hebrew," now housed in the Morgan Library, make up the single largest Native American archive in the nineteenth century.[34] In 1853, filling out the questionnaire and membership invitation from the Association for the Advancement of Science that asked about the "Branches of Science in which especial interest is felt," Thoreau gave this response: "The Manners and Customs of the Indians of the Algonquin Group previous to contact with the white men."[35] Native American languages, along with material artifacts, creation myths, and animal fables were constantly on his mind for the last fifteen years of his life.

I'd like to end with a passage, also from *Maine Woods,* that gives some sense of what it means to be so Indigenous-minded when the tables are momentarily turned, and it isn't the Native Americans but Thoreau himself who is the alien and misfit, when he suddenly finds himself surrounded by a pre-Columbian sonic habitat entirely without meaning for him:

> It was a purely wild and primitive American sound, as much as the Barking of a chickadee, and I could not understand a word of it . . . These Abenakis gossiped, laughed, and jested, in the

language in which Eliot's Indian Bible is written, the language which has been spoken in New England who shall say how long? These were the sounds that issued from the wigwams of this country before Columbus was born; They have not yet died away.[36]

Listening to the sounds of the Abenaki language, Thoreau suddenly has a kind of negative epiphany: these sounds belong to these woods, and he doesn't. Native Americans, of course, were the ones usually made to feel that way in the mid-nineteenth century. It says something about Thoreau that he is able momentarily to switch places with them, and subject himself to that surreal sense of involuntary nonbelonging. Doing so, he makes vulnerability a baseline human condition, the most widely shared, lowest common denominator for our species. Listening to Native Americans in that light, he hears them for what they are: veterans of survival and prophets in the wilderness, speaking in tongues that have "not yet died away" and perhaps will never die away, tongues whose vital import is just becoming apparent.

Nonextinct Sounds

In the twenty-first century those speaking tongues have only grown more eloquent. The soundtracks of now legendary names—Charley Patton, Jimi Hendrix, Mildred Bailey, Robbie Robertson, Link Wray, and Redbone—featured in Catherine Bainbridge's award-winning film, *Rumble: Indians Who Rocked the World,* make up one resurgent strand.[37] The live performances in the National Museum of the American Indian make up another.[38] A record number of Native Americans running for congressional and state legislative seats in 2018 add a third.[39] The growing number of Indigenous languages programs—from the two-year Master's program at MIT[40] to the eight language offerings at Yale;[41] to the pioneering American Indian Studies Research Institute at Indiana University, which uses digital tools and online "talking dictionaries" to offer instruction in Arikara, Assiniboine, and Pawnee to elementary schools, high schools, and community colleges—add yet another.[42] And the NEH-funded Standing Rock Lakota/Dakota Language Project at Sitting Bull College, which brings together the last generation of fluent speakers to transcribe Indigenous texts and make live recordings, suggests that the volume might be far from its upper limits.[43]

Kyle Powys Whyte, Potawatomi philosopher, author, and activist, is especially helpful at linking these resurgent sounds to time-tested strategies of adapting, honed through long practice. The massive disruptions that we now associate with climate change—ecosystem collapse, species loss, involuntary relocation, and pandemics—had always been part and parcel of New World colonialism, in play since the sixteenth century. Newly seen as apocalyptic in the twenty-first century, they seem far less so to Native peoples who have long suffered under them and lived through them, emerging with a different relation to the nonhuman world, and to ancestors and descendants both. "Climate adaptation" has always been key to Indigenous communities, Whyte argues, a form of knowledge that makes them veterans and pioneers in a "forward-looking framework of justice." Honoring "systems of responsibilities" ranging from "webs of interspecies relationships to government-to-government partnerships,"[44] Native Americans are now on the front lines of climate activism, setting into motion sounds of the future the rest of us are just beginning to rally to.

Native American reservations, taking up only 2 percent of the land in the United States, hold 20 percent of the nation's fossil fuel reserves, including coal, oil, and gas, worth some $1.5 trillion. Rather than privatizing and profiting from these reserves, Native Americans are among the most vocal opponents of a carbon-based economy. Deb Haaland, newly elected congresswoman from New Mexico, is committed to 100 percent renewable energy. "The fight for Native American rights is also a fight for climate justice," she said.[45] The National Caucus of Native American State Legislators, with eighty-one members from twenty-one states, has likewise made climate change and renewable energy one of its key advocacies.[46] Sheila Watt-Cloutier, chair for the Inuit Circumpolar Council and nominated for the Nobel Peace Prize in 2007, has pushed for the same agenda in global governance, working with Earthjustice and the Center for International Environmental Law to petition the Inter-American Commission to conduct hearings on the relation between climate change and human rights.[47]

This climate activism takes center stage in the Standing Rock Sioux Tribe's vigilance over the Dakota Access Pipeline, mobilized ever since the proposed plan was first announced in 2014.[48] Even though the pipeline was eventually approved by the Trump administration, this activism remains undiminished as Indigenous protestors monitor the oil leaks[49] while joining forces again in the opposition to the Keystone XL Pipeline,[50] and gearing up for a new fight against yet another proposal, known by the innocuous name "Line 3." The *Guardian* reports:

> Winona LaDuke, a veteran Native American activist and remarkable orator, has led a series of horseback rides along the pipeline route. Last year a group of Native youth organized a 250-mile "Paddle to Protect" canoe protest along the Mississippi River, which will be crossed twice by Line 3.
>
> If you want to hear what the resistance sounds like, "No Line 3" by Native rapper Thomas X is a good place to start; if you want to get a literal taste of it, Native women have routinely brought traditional breakfasts like frybread with blueberry sauce to the various public hearings over the project, sharing the food with everyone right down to the pipeline lawyers. (If you'd like you can also order some wild rice from LaDuke's Honor the Earth, one of the premier indigenous environmental organizations on the continent.)[51]

Here are the sights, sounds, and tastes of Indigeneity. Energized by crisis, it challenges the rest of us to be similarly resilient, to improvise in ways that keep extinction a live prospect and a deferred end, a catastrophe always before us and always receding, further and further into the distant future.

NOTES

1. Bernie Krause, *The Great Animal Orchestra: Finding the Origins of Music in the World's Wild Places* (Boston: Little Brown, 2012).

2. See http://www.wildsanctuary.com/.

3. Krause, *The Great Animal Orchestra*, 178.

4. Krause, 180–81.

5. Henry D. Thoreau, *Walden*, ed. Lyndon Shanley (Princeton, N.J.: Princeton University Press, 1971), 126.

6. Anthony Hallam and P. B. Wignall, *Mass Extinctions and Their Aftermaths* (Oxford: Oxford University Press, 1997), 1.

7. Richard Leakey and Roger Lewin, *The Sixth Extinction: Patterns of Life and the Future of Humankind* (New York: Anchor Books, 1996); Terry Glavin, *The Sixth Extinction: Journeys among the Lost and Left Behind* (New York: St. Martin's Press, 2007); Elizabeth Kolbert, "The Sixth Extinction? There Have Been Five Great Die-offs in History. This Time, the Cataclysm Is Us," *New Yorker,* May 25, 2009; Bill Marsh, "Are We in the Middle of a Sixth Mass Extinction?" *New York Times,* June 6, 2012.

8. S. L. Pimm, G. J. Russell, J. L. Gittleman, and T. M. Brooks, "The Future of Biodiversity," *Science* 269 (1995): 347–50; Edmund O. Wilson, *The Future of Life* (New York: Knopf, 2002), xxiii.

9. Elizabeth Kolbert, *The Sixth Extinction: An Unnatural History* (New York: Henry Holt, 2014), 17–18.

10. David B. Wake and Vance T. Vredenburg, "Colloquium Paper: Are We in the Midst of the Sixth Mass Extinction? A View from the World of Amphibians," *Proceedings of the National Academy of Sciences* 105 (2008): 11466–73.

11. PBS, "Frogs: A Thin Green Line," April 5, 2009, http://www.pbs.org/wnet/nature/episodes/frogs-the-thin-green-line/introduction/4763/.

12. John Upton, "Despite Deadly Fungus, Frog Imports Continue," *New York Times,* April 7, 2012, http://www.nytimes.com/2012/04/08/us/chytrid-fungus-in-frogs-threatens-amphibian-extinction.html.

13. Laura Gibbs, Introduction to *Aesop's Fables,* trans. Laura Gibbs (Oxford: Oxford University Press, 2002), xix.

14. These commonplace book entries from 1837 to 1847 are in the first volume of the *Journals,* ed. Bradford Torry and Francis Allen (Boston: Houghton Mifflin, 1962), 1:470.

15. For recent research on animal intelligence, see Frans De Waal, *Are We Smart Enough to Know How Smart Animals Are?* (New York: Norton, 2017); and Carl Safina, *Beyond Words: What Animals Think and Feel* (New York: Henry Holt, 2015). For an unforgettable account of the intelligence of parrots, see Allora and Calzadilla and Ted Chiang, "The Great Silence," *Planetary Computing,* May 8, 2015, http://supercommunity.e-flux.com/texts/the-great-silence/. My thanks to Kyle Hutzler for calling my attention to this piece.

16. Thoreau, *Journals,* 1:470.

17. Christopher Benfey, "The Lost Wolves of New England," *New York Review of Books,* January 22, 2013, http://www.nybooks.com/blogs/nyrblog/2013/jan/22/lost-wolves-new-england/.

18. "State Mammal List," Massachusetts Office of Energy and Environmental Affairs, http://www.mass.gov/eea/agencies/dfg/dfw/fish-wildlife-plants/state-mammal-list.html.

19. "Loons, Lead Sinkers and Jigs," Massachusetts Office of Energy and Environmental Affairs, http://www.mass.gov/eea/agencies/dfg/dfw/hunting-fishing-wildlife-watching/fishing/loons-lead-sinkers-and-jigs.html.

20. Stanley Cavell, *The Senses of Walden: An Expanded Edition* (Chicago: University of Chicago Press, 1992), 19–20.

21. Jeremiah 9:10.

22. Charles Darwin, *On the Origin of Species* (New York: New American Library, 1958), 108–9.

23. Thoreau, *Journals,* 8:221.

24. Henry David Thoreau, *The Maine Woods* (New York: Penguin, 1988), 6.

25. According to the Oxford English Dictionary, the adjective "extinct," referring to "that has died out or come to an end," first appeared in 1581. The word "extinction" appeared shortly thereafter in 1602.

26. Washington Irving, "Traits of Indian Character," in *The Sketch Book* (1820; New York: Signet, 1961), 273.

27. Irving, 280–81.

28. Irving, 282.

29. Irving, 282.

30. Herman Melville, *Moby-Dick* (1851; New York: Signet, 1951), 84.

31. Melville, 93.

32. Melville, 280.

33. Melville, 535.

34. Henry David Thoreau, *The Indians of Thoreau: Selections from the Indian Notebooks,* ed. Richard Fleck (Albuquerque: Hummingbird Press, 1974), 174.

35. Robert Sattelmeyer, *Thoreau's Reading: A Study in Intellectual History with Bibliographical Catalogue* (Princeton: Princeton University Press, 1988), 107.

36. Thoreau, *The Indians of Thoreau,* 185.

37. Joe Leydon, "Sundance Film Review, 'Rumble: The Indians Who Rocked the World,'" *Variety,* January 23, 2017, https://variety.com/2017/film/reviews/rumble-the-indians-who-rocked-the-world-review-1201966605/.

38. Videos of NMAI performances, https://www.si.edu/spotlight/native-american-music/videos-of-nmai-performances.

39. Ian Frazier, "Deb Haaland and Sharice Davids Lead a Record Number of Native American Candidates in the Midterms," *New Yorker,* November 5, 2018, https://www.newyorker.com/news/daily-comment/deb-haaland-and-sharice-davids-lead-a-record-number-of-native-american-candidates-in-the-midterms. Both Haaland, a member of the Pueblo of Laguna, and Davids, a member of the Ho-Chunk nation, were elected to the House of Representatives. Julie Turkewitz, "There's Never Been a Native Congresswoman. That Could Change in 2018," *New York Times,* March 19, 2018, https://www.nytimes.com/2018/03/19/us/native-american-woman-congress.html.

40. MIT Indigenous Languages Initiative, http://linguistics.mit.edu/mitili/.

41. "Native American Language Program offers 8 Indigenous Languages," https://ygsna.sites.yale.edu/news/native-american-language-program-offers-8-indigenous-languages-spring-2016-0. This program offers classes conducted via Skype with Cherokee, Choctaw, and Mohawk speakers in Oklahoma and Canada.

42. American Indian Studies Research Institute, https://www.indiana.edu /~aisri/projects/educational.html. In this context, see also the work of the National Breath of Life Archival Institute for Indigenous Languages at http:// nationalbreathoflife.org/.

43. Standing Rock Lakota/Dakota Language Project, https://humanitiesforall .org/projects/standing-rock-lakota-dakota-language-project.

44. See, for instance, Kyle Powys Whyte, "Justice Forward: Tribes, Climate Adaptation, and Responsibility," *Climate Change,* March 2013, https:// nwclimatescience.org/sites/default/files/2013bootcamp/readings/Whyte_2013b .pdf; "Indigenous Climate Change Studies: Indigenizing Futures, Decolonizing the Anthropocene," *English Language Notes* 55 (Fall 2017), 153–62 ; and "Indigenous Science (Fiction) for the Anthropocene: Ancestral Dystopias and Fantasies of Climate Change Crises," *Environment and Planning,* March 30, 2018, http://journals.sagepub.com/doi/10.1177/2514848618777621.

45. Tracey Osborne, "Native American Fighting Fossil Fuels," *Scientific American,* April 9, 2018, https://blogs.scientificamerican.com/voices/native-americans -fighting-fossil-fuels/.

46. National Caucus of Native American State Legislators, http://www.ncsl .org/research/state-tribal-institute/national-caucus-native-american-state -legislators.aspx.

47. Hari M. Osofsky, "Climate Change and Dispute Resolution Processes," in *International Law in the Era of Climate Change,* ed. Rosemary Gail Rayfuse and Shirley V. Scott (Cheltenham, UK: Elgar Edward, 2012), 353.

48. "Dakota Access Pipeline, What to Know about the Controversy," *Time,* October 26, 2017, http://time.com/4548566/dakota-access-pipeline-standing-rock -sioux/.

49. Matt Egan, "Dakota Access Pipeline Suffers a Minor Leak in April," *CNN Money,* May 10, 2017, http://money.cnn.com/2017/05/10/investing/dakota-access -pipeline-oil-spill/index.html.

50. "Rosebud Sioux Tribe Promises Continued Vigilance on the Keystone XL Pipeline," *BusinessWire,* November 21, 2017, https://www.businesswire.com/news /home/20171121005310/en/Rosebud-Sioux-Tribe-Promises-Continued-Vigilance -Keystone. On November 8, 2018, Judge Brian Morris of the District Court of Montana blocked construction of the Keystone XL pipeline by ordering the U.S. State Department to conduct a supplemental environmental impact review. Reuters, "U.S. to Conduct Additional Keystone XL Pipeline Review," https:// www.reuters.com/article/us-usa-keystone-pipeline/u-s-to-conduct-additional -keystone-xl-pipeline-review-idUSKCN1NZ2TH.

51. Bill McKibben, "Anti-pipeline Activists Are Fighting to Stop Line 3. Will they succeed?" https://www.theguardian.com/environment/commentisfree/2018 /jun/27/anti-pipeline-activists-fighting-to-stop-line-3.

CHAPTER 12

Hoopwalking
Human Rewilding and Anthropocene Chronotopes
PAUL MITCHELL

One method for reckoning with complex timescales is the chronotope (literally "time space"), developed by Mikhail Bahktin to denote the fusion of spatiotemporal indicators in novelistic discourse.[1] More generally, this concept is a means by which discourse can invoke historically configured and ordered tropes, a way to meaningfully situate actors and agency within a narrative. In so doing, chronotopes "invoke and enable a plot structure, characters or identities, and social and political worlds in which actions become dialogically meaningful, evaluated, and understandable in specific ways."[2] Any number of generic chronotopes—such as chronotopes of Western modernity in which human agency is "sovereign of nature," or chronotopes of romance, which situate agency in relation to the merger of the self with the desired other—might exist contemporaneously, since these are only devices for narrativizing action and agency.[3] With the category of the Anthropocene chronotope, I follow Dipesh Chakrabarty's historiographic claim about the Anthropocene that, as humans have become geological (i.e., evolutionary) agents recently in human history, "it is only very recently that the distinction between human and natural histories . . . has begun to collapse."[4]

An Anthropocene chronotope defines a genre, of which there are multiple manifestations, in which formerly reified categories of human (i.e., historical, anthropological) and natural (i.e., ecological, geological) timescales are imbricated.[5] While human life is always a part of natural time (such as seasons), the Anthropocene invites consideration and contextualization of human agency within timescales far beyond human scrutability. This sublimation of human agency and history into evolutionary, ecological, and planetary scales is emergent in discourse on science, politics, and society. As Vincent

Normand has articulated this new conceptual alignment, "once the window that separated us from the unlimited Nature 'outside' is broken, geology becomes a moral science (and vice versa)."[6] If the upshots of human agency have effects as durable as stratigraphy, with ramifications well beyond any terminal twig on the tree of life, then a further, wider reckoning of ethical horizons as a basis for judgment and action assumes a plausibility it may have not had before. Once paths between the present and deep futures are supposed, timescales formerly alien to quotidian experience may become both familiar and actionable, symbolically and practically. From a more myopic perspective, Anthropocene chronotopes call for ways of reckoning behavior that may appear irrational, inscrutable, or perverse.

A characteristic trope of the temporalizing modes designated as Anthropocene chronotopes is that the near future is, as a matter of practical and moral necessity, evacuated, overlain, and overdetermined by the *longue durée* of ecological time.[7] Whether this configuration represents something radically new or simply a lately rediscovered, forgotten sensibility, what is striking about Anthropocene chronotopes at the present is their increasing ubiquity and the plurality of their instances. Moreover, their reference to the master signification of ecological time serves to obscure the complex assemblage of discursive elements from other, all-too-human domains that ground lived and narrated experience in the Anthropocenic contemporary. "Rewilding," as an ecological and human praxis, is a rich, curious, and problematic Anthropocene chronotope, which I delve here through an ethnographic investigation in the American West, beginning with a genealogy of a concept. My purpose with this historical vignette is to trace the origins of the term "rewilding," not to follow the debate around the concept and its merits or deficiencies.

Protect, Restore, "Rewild": Deep Ecology to Deep Futures

"Wilderness protection" has been an explicit goal of Earth First! and other "eco-centric" North American radical environmental groups since their inception in the 1970s. These groups, born of the parallel developments of deep ecology and conservation biology, and the popularization of direct action tactics by anarchist writers, were perhaps most effective in making the conservation demands of mainstream environmental movements look comparatively moderate.[8] By the mid-1980s, with the academic emergence of restoration ecology and the recognition of "wilderness" as a conceptual mirage, another, more aggressive goal coalesced. In the September 1986 edition of the *Earth First! Journal*, ecology graduate student Reed Noss wrote that "Earth has been so ravaged by the pox of humanity" that "pristine wilderness" no longer exists. He proposed a "recipe for wilderness recovery," sketching a program for the creation of large areas of protected land scattered across bioregions, buffered against unwanted human and domestic animal intrusion, and connected by corridors to allow for animal and plant movement and gene flow. After the "presettlement reconstruction" of the bioregion, the reintroduction of large predators and missing species to these reserves would follow, supposedly re-creating ecosystems "like before humans destroyed [them]."[9]

The word "rewild" first appeared in print in a 1990 *Newsweek* article on the radical environmental movement reporting that an anonymous "eco-guerilla" vowed "not just to end pollution but to take back and 'rewild' one third of the United States."[10] By 1998, Noss and Michael Soulé, a conservation biologist, proposed rewilding as "the scientific argument for restoring big wilderness based on the regulatory role of large predators" in an ecosystem, claiming "aesthetic and moral" arguments for wilderness in the *Earth First! Journal*.[11] As presented in a 2005 *Nature* editorial, "Re-wilding North America," whose authors included Earth-First! cofounder Dave Foreman, rewilding is a project to "change the underlying premise of conservation biology from managing extinction" to restoring "some of the evolutionary and ecological potential that was lost" in North America upon human arrival in the late Pleistocene.[12] In effect, rewilding is proposed as a means to turn the tide in the "losing battle" of conservation, although the question of goalposts—i.e., of rewilding to what prior state—remains open.[13]

Although initial formulations of rewilding register as dated for silence on the effects of climate change in reconstructing prior ecosystem states, the concept of rewilding has diversified since its origins. More recent and ecumenical proposals define rewilding as the "passive management of ecological succession with the goal of restoring natural ecosystem processes and reducing human control of landscapes."[14] Prominent Harvard biologist E. O. Wilson's widely discussed proposal to conserve half the planet as the only means to ensure the preservation and restoration of biodiversity suggests that rewilding has become a distinctive and important, albeit controversial, aspect of restoration ecology.[15] Rewilding has been applied in Europe and North America through, for example, the introduction of multiple large herbivore species to the Netherland's Oostvaardersplassen reserve, or the re-introduction of gray wolves into Yellowstone.[16] These diverse experiments have met with varied bouts of success and praise, as well as failure and criticism.[17]

Rewilding has also been applied outside of wildlife ecology. In 2004, "re-wilding" appeared in the winter issue of *Green Anarchy* magazine as "a process of unlearning domestication" for the human species. One rewilder wrote that rewilding included "the prevention and undoing of social, physical, spiritual, mental, and environmental domestication and enslavement."[18] Human rewilding is set as contrast and antidote to the alienating, repressive, stultifying features of modern life, and variations on this term have carried through a wide range of texts in the last decade, from the writings of "anarcho-primitivist" authors to eco-psychology books.[19]

Recurrent in discourse on human rewilding is reference to mid- and late-twentieth-century anthropological literature, which tended to extol the cultures of foraging and stateless peoples and construe civilization, agriculture, and domestication as the defining tragedies of human history.[20] Of course, the "noble savage" trope often operative in this genre has a long pedigree: through the nineteenth-century Romantic movement, Rousseau, Dryden, and into antiquity. The gatherer-hunter as unalienated, egalitarian, vital, connected with nature, figures as an ideal of rewilding and is employed to critique modernity, providing a secular story of original sin in anthropological idiom, rooted in prehistory. Rewilding is now a common metonym for a variety of practices, from the paleo-diet to the profusion of "earth skills" or "primitive skills" gatherings in North America. The wings on the concept's metahistoric flight are somewhat clipped when presented as a commodifiable lifestyle fad for the affluent: nature therapy programs can "rewild your life"; increasing the diversity of the intestinal microbiome which is "rewild[ing] your gut."[21]

Hoopwalking

While now profuse in its denotations, rewilding remains a potent concept in radical environmental politics and activism—perhaps especially so in queer activism. While critical studies of gender and environment have led to burgeoning academic discourse of queer ecology, these discussions have been largely absent in the consideration of rewilding.[22] In contrast, the fall 2015 issue of *Radical Fairy Droppings*—the periodical produced by a radical fairy collective, a part of a worldwide movement for alternative queer consciousness begun in the United States in the 1970s—was devoted to rewilding. In its pages, alongside poems and paeans to old-growth forests, and recipes for wild foods (e.g., "Two Mouse Stew" with wild roots), Finisia Medrano published a "Letter of Intention and Invitation" to readers to join a "traditional, symbiotic life way that was practiced here in the Columbia and Snake River Plateau for tens of thousands of years, before its historic genocide."[23]

Finisia is white, about sixty years old, and has lived in the western United States most of those years.[24] In her 2011 book *Growing Up in Occupied America*, Finisia recounts her life as a teenage runaway and itinerant Christian evangelist living out of a horse-drawn wagon, an artist and wanderer in the American West for decades. It was during that time, she says, that she learned "old Indigenous ways" from Shoshone elders. She is now a nodal figure in the High Desert Wildtending Network, a loose network of perhaps a few hundred who claim dedication "to the regeneration and sustenance of Indigenous Migratory Life-Ways through tending Wild Food Gardens" in the Great Basin and Columbia Plateau bioregions of the western United States. The group's name comes from ecologist M. Kat Anderson's 2005 book on practices of plant and landscape management, *Tending the Wild*, which describes landscape and plant management practices among Indigenous people in California. This "wild tending" created gardens of wild foods, and dispels the myth that landscapes presumed pristine wilderness by Euro-Americans were devoid of formative human influence before European presence, in California and elsewhere.[25]

In summer 2016, I traveled with a few dozen in this network around the corners of Oregon, Idaho, and Nevada. Throughout the year, these rewilders gather and replant native roots, fruits, berries, nuts, and greens. In the winter, some tan hides for use, gifting, or sale. They follow a seasonal cycle of movement on "the hoop," referring to a circuit of wild food gardens in the region tended for subsistence continuously by Indigenous people until much of the last century. The referents of the hoop include the long-cultivated gardens of seasonal wild food spread around the bioregion, the practice of living on and tending these gardens, and the spiritual and social significance of the "sacred hoop" or "circle of life."[26] The hoop is described as a relational ethic. Finisia defined it this way: "I don't take a berry anywhere that I don't think about that plant, how it's desiring to reproduce, how I can get those seeds back in the ground." It means that not all the seeds go back, but some do. It means that plants, animals, and landscapes are to be "given life back for the life they give."

Those in the High Desert Wildtending Network are "hoopwalkers."[27] Although the hoopwalkers I met estimated that they ate, depending on the season and circumstance, at most about a third of their diet in wild foods, their primary interest is not subsistence but rather to increase the abundance of the plants and knowledge about precolonial foodways. They work on both public and private land, wherever gardens are safe from cattle grazing, heavy human traffic, and development. Very few have lived full time "on the hoop" over the years; most shift in and out seasonally, on the weekends, or between odd jobs. Some rent, some live in cars, some live in trailers. Only Finisia and a few of her students, all white, have lived through the year traveling on horses.

Many hoopwalkers identified as queer or had felt misfit growing up in conservative rural America. I suspect that I had relatively easy access and rapport with them because I am a cis white male from rural Midwest America. One hoopwalker suggested that they were "people who are already outsiders. [We] don't fit into that structured system." In the decade or so that Finisia has been actively recruiting from radical fairy sanctuaries and anywhere else, hundreds have lived with and learned from her for days, weeks, or years at a time. Finisia does not charge money, but students should "be generous" and "teachable."[28] Although Fin is notoriously abrasive— within minutes of meeting me, she called me an "ecocidal motherfucker" for driving a car—she wins respect and students easily. She hopes, through teaching and planting, to work toward increasing the abundance of wild food plants until there is sufficient density to support humans living part or full time as foragers, away from "civ" or "Babylon," forging a lifeway resilient to the social and ecological collapse hoopwalkers sense looming.

While I was with Finisia, it was the season to collect yampa (in Shoshoni; wild carrots, *Perideridia spp.*) and huckleberries (*Vaccinium spp.*). At her request, I drove her down old logging roads toward yampa meadows. She looked out the window and up the hillside to greet blackcap raspberry (*Rubus leucodermis*) bushes that she had planted years before. She exclaimed to plants dotted sporadically down the slope, "Wild food! This is a gastric religion!"

Within hours of meeting her, Finisia invited me to live on the hoop. I hesitated. She shook her head, said that even she had things she needed to finish "in civ" before leaving. "I had to transition," she laughed. "Get your stupid PhD and then get your dumb ass out on the hoop."

Transition

Finisia was not born a woman. She lived as a gay man for years. A wealthy boyfriend paid for her gender confirmation surgery in the late 1970s.[29] After she separated from him, she found herself, as she put it, among "drunken Christian Indians," the only ones who would accept her as a transwoman.[30] She talks openly about transitioning from man to woman, about learning how to perform femininity. She calls herself "Tranny Granny" and says she's good for the hoop, relishing that "the gossip mill runs wild when word gets around that a transsexual is riding around on horseback in the scrubland, planting wild carrots."[31]

"Transition," both as word and concept, suffuses the hoop. I frequently heard the word in conversation with hoopwalkers. Working toward replanting "human habitat" and forming community such that kids can grow up on the hoop is a common goal. When I asked about some things around camp that might be at variance with that ideal, like cars or cell phones, or the social media sites maintained by hoopwalkers, I was told that "being a *transition person* is hard," but it isn't necessarily a problem to use cars or phones or the Internet, so long as these means serve their hoopwalking. Fin said, "I know why I'm in the car, I know why I'm with you. I'm gonna have that hamburger because it's gonna power my little butt out there . . . I turn it into hoop!" (Since the summer I spent with hoopwalkers described here, Finisia has been forced to give up horseback life due to knee problems, and has acquired a truck to "motorize" her hoopwalking.)

All the hoopwalkers I met know that it's not possible to live wholly apart from what they are trying to escape. Many collect food stamps and work odd jobs to keep planting wild foods at least some of the year. But, being a transition person is not only a matter of having the material residues of the mass market adhere; the conditioning of being born and raised in "Babylon" is psychically indelible. Lycos[32] had been a schoolteacher for much of his life, a hoopwalker for a decade. He said: "You know, you can't completely shed your double-mindedness. You bring your children up out here, they might be able to. But as far as the ones who are coming now, they're just coming to learn the skills. If they can accumulate skills, that will help the generations afterwards."

For Lycos, the project of transition is not only that of learning the skills to live on the hoop, but that of losing one's double-mindedness, an effect of the psychological conditioning accrued by life "in civilization." Double-mindedness, a biblical term often on Finisia's lips, is, for her and her fellow travelers, "lack of purified intention," a liminal and split psychic state of the person in transition to life on the hoop. That transition is always incomplete, not just because one is caught in the material realities of the present but also because of the durable psychic impress of the lifeworld "of Babylon" on the self.

The hoopwalkers understand their work on the hoop as work on the self in transition. Many talked about addiction and trauma in their earlier lives, and healing through hoopwalking. Lycos told me that, in the future, songs will be sung about "how hard it was to be a transition person—cause it's the hardest thing. It's easy being in Babylon!" He described himself as part of a "transition generation," perhaps the first of many, whose descendants "in spirit" will be able to do what a "transition person" cannot: live fully on the hoop, free from the conditioning of Babylon.[33] Lycos suggested that it was his conditioning that prevented stable, enduring community, what he called "clan," on the hoop: "And this might be part of the burden of being transition people, because we may never, never be able to be clan, but we will do the work that will enable the people who will follow to be clan, because we will have worked through a lot of our shit." "Clan" is fictive kinship and a normative goal, a social ideal possible when transition is complete.[34]

That "transition" is a circulating term and metaphor for the personal, social, and inter-generational journeys of the hoopwalkers is not surprising, given Finisia's influence and the presence of other trans people on the hoop. The metaphor of transition suggests both an embodied process of personal transformation, but also a performance, a habitus appropriate to an achieved, rather than an assigned, category. For the Hoopwalkers, transition is not movement toward the natural but rather toward a careful artifice. They employ chronotopic terms of transition in multiple imbricated domains, implicating transitions of the individual, social, generational, and ecological into each other. The hardships, personal and social, of the "transition person" are justified within the place of the individual in a "transition generation." These metaphorical lines moor hoopwalking discourse to some recent concerns of queer and trans ecology, but the durative temporal mode of transition competes with another, teleological and terminal metaphor on the hoop, that of "return."[35]

The Return?

On August 11, 2016, the United States National Park Service (NPS) approved a rule on the "Gathering of Certain Plants or Plant Parts by Federally Recognized Indian Tribes for Traditional Purposes," which "established a management framework" to allow certain members of federally recognized tribes to gather plants for "traditional purposes."[36] Citing many of the same works on Indigenous ecological practices that hoopwalkers recommended to me, the report states: "Research has shown that traditional gathering, when done with traditional methods . . . and in traditionally customary quantities, may help to conserve plant communities."[37] Prior to this rule, the NPS did not allow plant gathering by tribes except in the case of specific laws or treaties.[38]

The majority of the ninety public comments on the rule as proposed in 2015 were submitted by federally recognized tribes and other Indigenous groups and individuals, and a number of these suggested that the proposed rule did not adequately protect information about Indigenous sacred sites, plants, or practices. In a July 29, 2015, comment, the Tribal Environmental Policy Center wrote: "Many Tribes strive to keep their traditional practices hidden from those outside their respective communities to prevent overgathering or damage by others, or for other cultural reasons." This comment and others like it did not substantially change the final version of the new park rule, but it draws attention to the contestations over the skills and knowledge that hoopwalkers practice. These contestations settle on the question of to whom a lifeway belongs.

In 2014, photographer Adrian Chesser collaborated with Native writer and ritualist Timothy White Eagle on a photographic and essay volume on the hoopwalkers, *The Return*. All the hoopwalkers I met are at pains to express that they know they are not returning to the lifeways of their mostly European ancestors. In "returning to the old ways," the "return" is not a reversion but rather a process of experimenting toward a lifeway that does not further entrench ruinations of colonialism, capitalism, and intensive monocrop agriculture in the American West. Hoopwalkers articulated return primarily in relation to the landscape and ecosystem—hoopwalking is how to live symbiotically "*here* in the Great Basin and Columbia and Snake River Plateau," with its native plants—but did not claim to fix it to only one point in time or bounded, genealogical *ethnos*.[39] As Indigenous scholars question whether so-called Indigenous knowledge should be shared or used outside of the context of sustaining Indigenous peoples against continuing settler oppression, white hoopwalkers claim that the "symbiotic lifeway" between humans and a particular landscape that came before can come again.[40]

The relation of that claim to the politics and stakes of Indigeneity is unresolved among hoopwalkers.[41] The hoopwalkers are not simply mimetic of Indigenous cultures: their language and apocalyptic sensibility are steeped in Judeo-Christian tradition. Hoopwalkers reference Indigenous lifeways as a salvific practice, a response to their inherited original sin. As Lycos said, it was "the whites who broke the hoop, so they should help to fix it." Others told me they were concerned about cultural appropriation, but that the skills should be open to anyone who was committed, respectful, and an ally to Native

Americans and their struggles. Was this more than a liberal gesture of symbolic recognition, a salve to continuing practices of white settler dispossession? When I asked Finisia why she wore beaded buckskin (a kind of dress usually reserved for ceremonial occasions), she snapped back: "This isn't just about *drag*, about going around *dressed up like an Indian!* Either you are planting back [i.e., hoopwalking] or you are just another part of the Swallowing Monster."

Although most hoopwalking goes on without explicit consultation with local tribal communities, a few hoopwalkers told me they wished for more involvement with Native people. The only Native person I met while on the hoop was a Modoc man from southeastern Oregon, Jack, who allowed Fin to graze horses on his property in the past. Jack was a friend to some hoopwalkers, but didn't go planting. He leaned against his truck as I asked him why. "Oh, the difference between me and them is that they [the hoopwalkers] are waiting for the apocalypse. I think it already happened."

Indigenous scholars like Kyle Powys Whyte have noted that Indigenous peoples in settler colonial societies "already inhabit what our ancestors would have understood as a dystopian future."[42] For these ancestors, this world may appear to be a kind of horrific fantasy. Certainly a kind of fantasy is at work in the assumption, central to many human rewilders, that Indigenous people and their lifeways can be, in effect, slotted as those of the "last people living in Holocene conditions," before the Anthropocene.[43]

Rewilding, Hoopwalking, and Anthropocene Chronotopes

Both ecosystemic and human rewilding can be understood as Anthropocene chronotopes. Rewilding originated in radical environmentalism, the eco-centrism of which invokes temporalities beyond the human, making human action accountable to times and spaces beyond immediate sensibility. Rewilding has accrued multiple, variegated meanings, some of which are contradictory, from carving out spaces to restore to a prior state of ecosystem functioning, to allowing natural processes to resume, or to reducing human management of landscapes. All these modalities of reassembling the human in a larger, temporally more expansive lifeworld imbricate human and ecological time.[44] Rewilding necessarily affirms what environmental historians have long intoned: that wilderness is an illusion, that no ecosystem is free from human influence.[45] No "rewilderness" can be, either. Fitting human agency into ecological time in a manner that preserves, creates, and recreates wildness is, as one rewilder and occasional fellow traveler with the hoopwalkers put it, "trying to figure out how humans fit into the ecosystem without destroying it."[46]

Hoopwalking is an experiment in conjoined ecosystemic and human projects of rewilding. The hoopwalkers metaphorically align personal and ecological temporal scales by situating biographical transitions from old to new selves and an intergenerational transition from "Babylon" to the hoop within a narrative of return to "symbiotic lifeways" in the American West. In so doing, they situate their action in frames of time and space at once immediate and durative ("transition") as well as epochal and teleological ("return"). Thus, tending the rewilderness is not only about making "human habitat" for future, it is about inhabiting canyons, creek bottoms, scrublands, and pine forests today, and sowing selves and relations there. But political, not only temporal, tension pervades this layered chronotopic alignment.

Hoopwalking is at once a settler project overlain with Judeo-Christian motifs and stereotypically American imaginaries about the meaning of nature, as well as a project to sustain forms of knowledge, practice, and relationship with landscapes that have been pushed near to extinction through logics of settler colonialism and capitalism. Although refusing to accept a chimerical vision of "wilderness" devoid of humans, hoopwalking as return may posit lifeways out of time. Despite valuing Indigenous knowledge as potentially salvific, hoopwalking does not clearly answer questions raised about its appropriation of Indigenous knowledge and practice. As an experimental prefiguration of a positive place for humans in a restorative ecology, hoopwalking proposes deep futures reflecting particular lived histories in the pasts of specific places, but it is deeply entrenched in present relations of power and privilege.

From a wider vantage, while the long past is perhaps most commonly invoked to naturalize the present, the Anthropocene can be used to call upon this timeframe as a basis for ethically framing human agency within ecosystems for the long haul.[47] But not, perhaps, without its contradictions: whatever its restorative aspects,

Anthropocene rhetoric can be a device for the naturalization of human dominance, enshrining instrumental reason and the white male *anthropos* in a theological narrative of dominance over nature.[48] The ethical ambiguity of the concept draws into question the natures and the practices marshalled under its sign. As scholars have asked which *anthropos* is the actor in the Anthropocene, interest in Anthropocene chronotopes should attend to not only the temporal (*chronos*) but also the spatial (*topos*) of this analytic. The time of the Anthropocene does not manifest the same everywhere: preparation for collapse for some is life after apocalypse for others.

In sharing some of what they have shared with me, I cannot claim to speak for the hoopwalkers. Drawing attention to their project brings to the fore perhaps a limit case of the Anthropocene chronotope, a way of being that no doubt many would find absurd, unlivable, or unconscionable. It also lays bare a core tension within the Anthropocene concept: that fixing politics at the level of the species obscures human differences of power, agency, and responsibility.[49] These political categories do not disappear upon the invocation of "Anthropocenic-chronotopic" scales, such as transition or return on the hoop: they manifest discursively and materially in questions of who can transition, and into what, and who can return, and to where. More broadly, one must ask how the political ecology of the American West, the dispossession of Native land in the enduring structure of settler colonialism, and the social and demographic transitions of rural America in recent decades shape this narrative world. The High Desert Wildtending Network exists within and because of these realities and is emergent from them as much as it is from radical queer consciousness or precolonial lifeway practices. The Anthropocene chronotopes that undergird such projects may not only flatten but also entrench the structures that have, thus far, meted out apocalypse differentially.

Coda

Bruce Braun has characterized rewilding as a project of "making new natures" that "is not just artifice, it is experimental all the way down."[50] Experiments are explorations in uncertainty, and many of the hoopwalkers with whom I spoke are unsure, even gloomy, about their success, however it may be measured. Some transitions may never be complete, some returns may fall short. Finisia, hyperaware of ambient and accelerating existential dangers, a doomsday prophetess of nuclear warheads and rising global temperatures, wrote about being "pretty sure we're gonna see the end of all things. And that's not just civ things. That's all things." Why continue, then? "It's this attempt to, in real time performance art, fulfill every hopeful promise of return and renewal, every vision and prophecy of such a thing. Do I believe it? Fuck no. Do I love that shit? Absolutely, I'm gonna to do it with all my heart."[51]

Some of Finisia's acquaintances, a local retiree couple who come by occasionally to pick huckleberries with Fin when she's around, pulled into camp after I started recording a conversation. We were seated, and Fin told them to roll up some logs. Before I could introduce myself, she announced, "This is Paul. He's an anthropologist. He's out here. He's out to gather ammunition for the guys that are shooting at me, and I'm here to help him to fill his pockets." She laughed, and I did, too, despite myself. Before I could decide how to respond, Finisa turned the conversation to berries, their colors, tastes, distribution around camp that time of year.

Finisia gave me plenty of ammunition. As if she welcomed the shots to come, she answered any question put to her. On the morning I left, she hugged me and gave me the quart of huckleberries she had picked the previous day. "Be sure to shit out those seeds somewhere they'll grow."

NOTES

1. Mikhail Bahktin, *The Dialogic Imagination,* trans. Caryl Emerson and Michael Holquist (Austin: University of Texas Press, 1981).

2. Jan Blommaert, "Chronotopes, Scales, and Complexity in the Study of Language in Society," *Annual Review of Anthropology* 44 (2015): 105–16.

3. David Lipset, "On the Bridge: Class and the Chronotope of Romance in an American Love Story," *Anthropology Quarterly* 88, no. 1 (2015): 165–86.

4. Dipesh Chakrabarty, "The Climate of History: Four Theses," *Critical Inquiry* 35, no. 2 (2009): 207.

5. Kate Wright, *Transdisciplinary Journeys in the Anthropocene: More-than-Human Encounters* (Oxford: Routledge, 2017), 201; David Lipset, *Yabar: The Alienations of Murik Men in a Papua New Guinea Modernity* (New York: Springer,

2017); Mary Louise Pratt, "Coda: Concept and Chronotope," in *Arts of Living on a Damaged Planet,* ed. Anna Lowenhaupt Tsing, Nils Bubandt, Elaine Gan, and Heather Anne Swanson, 169–75 (Minneapolis: University of Minnesota Press, 2017).

6. Vincent Normand, "In the Planetarium: The Modern Museum on the Anthropocenic Stage," in *Art in the Anthropocene,* ed. Heather Davis and Etienne Turpin, 63–78 (London: Open Humanities Press, 2015).

7. See, for example, Joel Wainright and Geoff Mann, "Climate Leviathan," *Antipode* 45, no. 1 (2013): 1–22.

8. Christopher Manes, *Green Rage: Radical Environmentalism and the Unmaking of Civilization* (Boston: Little, Brown and Company, 1990); Bron Taylor, "Three Tributaries of Radical Environmentalism," *Journal for the Study of Radicalism* 2, no. 1 (2008): 27–61; Jonathan Hintz, "Some Political Problems for Rewilding Nature," *Ethics, Place, and Environment* 10, no. 2 (2007):177–216.

9. Reed Noss, "Recipe for Wilderness Recovery," *Earth First! Journal* 6, no. 8 (1986): 22–25.

10. Jennifer Foote, "Trying to Take Back the Planet," *Newsweek* 115, no. 6 (1990): 24.

11. Michael Soulé and Reed Noss, "Rewilding and Biodiversity: Complementary Goals for Continental Conservation," *Wild Earth* 8 (1998): 19–28.

12. Josh Donlan, Harry W. Greene, Joel Berger, Carl E. Bock, Jane H. Bock, David A. Burney, James A. Estes, Dave Foreman, Paul S. Martin, Gary W. Roemer, Felisa A. Smith, and Michael E. Soulé. "Re-wilding North America," *Nature* 436 (2005): 913.

13. Harry Greene, personal communication, March 20, 2016.

14. Dolly Jørgenson, "Rethinking Rewilding," *Geoforum* 65 (2014): 482–88; David Nogués-Bravo, Daniel Simberloff, Carsten Rahbek, and Nathan James Sanders, "Rewilding Is the New Pandora's Box in Conservation," *Current Biology* 26 (2016): 83–101; L. Gillson, R. J. Ladle, and M. B. Araújo, "Baselines, Patterns and Process," in *Conservation Biogeography,* ed. R. J. Ladle and R. J. Whittaker, 31–44 (Oxford: Wiley-Blackwell, 2011).

15. E. O. Wilson, *Half Earth: Our Planet's Fight for Life* (New York: Liveright, 2016).

16. Elizabeth Kolbert, "Recall of the Wild," *New Yorker,* December 24 & 31, 2012; Jamie Lorimer and Clemens Driessen, "Wild Experiments at the Oostvaardersplassen: Rethinking Environmentalism in the Anthropocene," *Transactions of the Institute of British Geographers* 39, no. 2 (2014): 169–81; Douglas W. Smith, Rolf O. Peterson, and Douglas B. Houston, "Yellowstone after Wolves," *BioScience* 53, no. 4 (2003):330–40.

17. JoAnna Klein, "'Rewilding' Missing Carnivores May Help Restore Some Landscapes," *The New York Times,* March 16, 2018; Patrick Barkham, "Dutch Rewilding Experiment Sparks Backlash as Thousands of Animals Starve," *The*

Guardian, April 27, 2018; T. Caro, "The Pleistocene Re-wilding Gambit," *Trends in Ecology and Evolution* 22, no. 6 (2007): 281–83; and H. N. Huynh, "Pleistocene Re-wilding Is Unsound Conservation Practice," *BioEssays* 33, no. 2 (2010): 100–102.

18. Peter Michael Bauer, "Introducing . . . Rewilding," http://www .petermichaelbauer.com/introducing-rewilding/ (accessed May 1, 2017).

19. John Zerzan, *Future Primitive Revisited* (Port Townsend, Wash.: Feral House, 2012); Peter H. Kahn Jr. and Patricia H. Hasbach, "The Rewilding of the Human Species," in *The Rediscovery of the Wild*, ed. P. H. Kahn Jr. and P. H. Hasbach (Cambridge, Mass.: MIT Press, 2013), 207–32.

20. For example, Stanley Diamond, *In Search of the Primitive: A Critique of Civilization* (Brunswick, N.J.: Transaction, 1974); Colin Turnbull, "Cultural Loss Can Foreshadow Human Extinctions: The Influence of Modern Civilization," in *Extinctions*, ed. M. Nitecki (Chicago: University of Chicago Press, 1984), 175–92; Pierre Clastres, *Society against the State*, trans. Robert Hurley (New York City: Zone Books, 1987).

21. We Are Wilderness, "Rewild Your Life 30 Day Challenge," http://learn .wearewildness.com/p/rewildyourlife (accessed May 1, 2017); Robyn Chutkan, "Rewild Your Gut," *Omega*, May 20, 2016, https://www.eomega.org/article/rewild -your-gut-microbiome.

22. Timothy Morton, "Queer Ecology," *PMLA* 125, no. 2 (2010): 273–82; Catriona Sandilands, "Lesbian Separatist Communities and the Experience of Nature: Toward a Queer Ecology," *Organization and Environment* 15, no. 2 (2002): 131–63; Catriona Sandilands and Bruce Erickson, eds., *Queer Ecologies: Sex, Nature, Politics, Desire* (Indianapolis: Indiana University Press, 2010); Kelly Struthers Montford and Chloë Taylor, "Feral Theory," *Feral Feminisms* 6 (2016): 5–17.

23. Finisia Medrano, Walker, Seda, and Neisan, "Interview with Finisia Medrano," *Radical Feral Droppings* 41, no. 3 (2015): 21.

24. Finisia Medrano passed away on April 3, 2020.

25. Quote from the High Desert Wildtending network's Facebook page, https://www.facebook.com/wildtendingnetwork/ (accessed May 1, 2017); M. Kat Anderson, *Tending the Wild: Native American Knowledge and the Management of California's Natural Resources* (Berkeley: University of California Press, 2005).

26. Dennis W. Zotigh, "History of the Modern Hoop Dance," *Indian Country Today*, May 7, 2007, https://indiancountrymedianetwork.com/news/history-of -the-modern-hoop-dance/; John G. Neihardt, *Black Elk Speaks: The Complete Edition* (Lincoln: University of Nebraska Press, 2008); Paula Gunn Allen, *The Sacred Hoop: Recovering the Feminine in American Indian Traditions* (Boston: Beacon Press, 1992).

27. For the sake of consistency, I use "hoopwalker" in this text. *Hoopwalker* and *hoopster* are both common self-designations in the High Desert Wildtend-

ing Network, although not everyone uses or accepts "rewilder." Prior to 2016, "Sacred Hoop Rewilding" was the name of the educational nonprofit organization run by some members of the network that was changed to High Desert Wildtending Network.

28. Medrano et al., "Interview with Finisia Medrano," 30.

29. Finisia Medrano, *Growing Up in Occupied America* (Raleigh, N.C.: Lulu Press, 2011), 22–24.

30. Medrano et al., "Interview with Finisia Medrano," 25.

31. Medrano et al., 29.

32. Names to which direct quotes are attributed in this article are pseudonyms, except for Finisia Medrano, who consented to identification.

33. See "spiraling temporality" in Kyle Whyte, "Indigenous Science (Fiction) for the Anthropocene: Ancestral Dystopias and Fantasies of Climate Change Crises," *Environment and Planning E: Nature and Space* 1 (2018): 224–42.

34. See nonheteronormative futurity in Nicole Seymour, *Strange Natures: Futurity, Empathy, and the Queer Ecological Imagination* (Champaign: University of Illinois Press, 2013); Donna Haraway, "Anthropocene, Capitalocene, Plantationocene, Chthulucene: Making Kin," *Environmental Humanities* 6 (2015): 159–65.

35. Nicole Seymour, "Transgender Environments," in *Routledge Handbook of Gender and Environment,* ed. Sherilyn Macgregor (New York: Routledge 2017), 253–69.

36. National Park Service, "Gathering of Certain Plants or Plant Parts by Federally Recognized Indian Tribes for Traditional Purposes," July 12, 2016, https://www.federalregister.gov/documents/2016/07/12/2016-16434/gathering -of-certain-plants-or-plant-parts-by-federally-recognized-indian-tribes-for -traditional.

37. Douglas E. Deur and Nancy Turner, *Keep It Living: Traditions of Plant Use and Cultivation on the Northwest Coast of North America* (Seattle: University of Washington Press, 2005); B. Parlee and F. Berkes, "Indigenous Knowledge of Ecological Variability and Commons Management: A Case Study on Berry Harvesting from Northern Canada," *Human Ecology* 34 (2006): 515–28.

38. Michelle Tirado, "National Park Service Does Face-Plant with New Rule on Gathering Plants," *Indian Country Today,* August 20, 2015, https:// indiancountrymedianetwork.com/news/politics/national-park-service-does-face -plant-with-new-rule-on-gathering-plants/; Kollibri Terre Sonnenblume, "A Century of Theft from Indians by the National Park Service," *Counterpunch,* March 9, 2016, http://www.counterpunch.org/2016/03/09/a-century-of-theft -from-indians-by-the-national-park-service/.

39. Medrano et al., "Interview with Finisia Medrano," 21.

40. Kyle Whyte, Chris Caldwell, and Marie Schaefer, "Indigenous Lessons about Sustainability Are Not Just for 'All Humanity,'" in *Situating Sustainability:*

Sciences/Arts/ Societies, Scales and Social Justice, ed. Julie Sze, 149–79 (New York: NYU Press, 2018).

41. Bruno Seraphin, "'Paiutes and Shoshone Would Be Killed for This': Whiteness, Rewilding, and the Malheur Occupation," *Western Folklore* 76, no. 4 (2017): 447–78. See also Scott Lauria Morgensen, *Spaces Between Us: Queer Settler Colonialism and Indigenous Decolonization* (Minneapolis: University of Minnesota Press, 2011), especially 127–94.

42. Kyle Powys Whyte, "Our Ancestors' Dystopia Now: Indigenous Conservation and the Anthropocene," in *Routledge Companion to the Environmental Humanities,* ed. Ursula Heise, Jon Christensen, and Michelle Niemann, 206–15 (New York: Routledge, 2017); Bruno Seraphin, "'The Hoop' and Settler Apocalypse," *The Trumpeter* 32, no. 2 (2016): 126–46.

43. Gregory D. Smithers, "Beyond the 'Ecological Indian': Environmental Politics and Traditional Ecological Knowledge in Modern North America," *Environmental History* 20 (2015): 83–111.

44. Gillson, Ladle, and Araújo, "Baselines, Patterns and Process."

45. William Cronon, "The Trouble with Wilderness: Or, Getting Back to the Wrong Nature," *Environmental History* 1, no. 1 (1996): 7–28.

46. Peter Michael Bauer, in interview with Monica Weitzel, *Community Hotline,* "Community Hotline 3/8/17 Rewild Portland," https://www.youtube.com/watch?v=vhoIXGVdC-U (accessed May 01, 2017).

47. Bruno Latour, "Agency at the Time of the Anthropocene," *New Literary History* 45 (2014): 1–18.

48. Lauren A. Rickards, "Metaphor and the Anthropocene: Presenting Humans as a Geological Force," *Geographical Research* 53, no. 3 (2015): 280–87; Jeremy Baskin, "Paradigm Dressed as Epoch: The Ideology of the Anthropocene," *Environmental Values* 24 (2015): 9–29.

49. Andreas Malm and Alf Hornberg, "The Geology of Mankind? A Critique of the Anthropocene Narrative," *The Anthropocene Review* 1, no. 1 (2014): 62–69.

50. Bruce Braun, "From Critique to Experiment? Rethinking Political Ecology for the Anthropocene," in *The Routledge Handbook of Political Ecology,* ed. Tom Perreault, Gavin Bridge, and James McCarthy (London: Routledge, 2015), 102–14.

51. Medrano et al., "Interview with Finisia Medrano," 26.

CHAPTER 13

Dirt Eating in the Disaster

IEMANJÁ BROWN

The mouth is a good place for collapsing human and geological time-scales. On the one hand, anxiety about ongoing and immanent climate disaster lends itself to a singular conception of the human who is all mouth, devouring what used to be called nature, annihilating it and thus also self-destructing.[1] But appetite, as an urge to get close to something by incorporating it into the body, is also productive of numerous relations other than those of destruction. The attempt to conceive of climate change requires thinking at once across multiple scales. In other words, maybe the desire to ingest can be an urge to sense the unthinkable by some other means. The literal appetite for dirt is but one site for this alternate relation between human and milieu that can jump scales. Craving dirt opens the human body up to becoming a small, temporary vessel for the varied social and material histories that are sedimented in ground.[2]

Geophagy, or dirt eating, is listed in the *Diagnostic and Statistical Manual of Mental Disorders* under "pica," which is the desire to eat "non-nutritive substances."[3] In the United States, the pathologized practice is usually associated with the south, particularly with Black women, poor women, and pregnant people of all races. Since geophagy has historically been marked as abject, it has dirty connotations that land unevenly along raced, classed, and gendered humans. Even though a geologist might prefer to call the geophagist a clay eater, I stick to the word *dirt* because the ways dirt can't shake the dirty are also constitutive of what comes together in the geophagist's mouth.

But what comes into the geophagist's body is also the result of a pursuit of pleasure, a pursuit that tends to be most urgent after it has rained. One moment of this kind of urgency arose for me in New Jersey. On my way home to New York City, I passed by the familiar scene of sad urban appendage: a port that's been mechanized and moved to the city's periphery and an oil refinery with clean energy signs and plumes of smoke majestically saying otherwise. Feeling the guilt of being a human who is deeply entangled with infrastructures that merge into the background as they devour, I caught sight of a construction site next to the highway. The area was empty except for a backhoe and a pile of red earth clumped together in a patterned mess. The pile was the kind of red that tugs at me and expands my mouth, as if my insides are making room for what they've been missing. This is not a confession, but a description of a small desire—one that is not about eating as a will toward survival, but about appetite as a will toward pleasure.

After seeing the small, upturned pile of dirt, I began poring over soil maps of New Jersey. I found that one possible origin for the color is that it emerges from red shale, which is most common in tropical regions but exists in some parts of the north. It tends to form in temperate and humid environments, weathering rapidly to build up layers of the kind of red clay that is often associated with the identity of the Deep South (see the abundance of music, poetry, novels, memoirs with "red dirt" in the title). The red shale bedrock of northeast New Jersey is considered young, formed in the warm weather of the Triassic and Jurassic periods 150–250 million years ago. When the last glacier receded from that area about 20,000 years ago, creating New York City and Long Island with its moraine, it helped distribute some of the shale to the soil's surface. So the red is young for a geologist, old for the rest of us, and appetizing to some of us. And the appetite for that color is one that, I think, senses how the localized pile of upturned dirt is stratified with a vast temporal expanse. Dirt eating thus makes the human organism a container for a timescale that precedes and outlasts, if not the species, the ways of being human with which we are familiar.

That red ground in New Jersey is also a product of an elaborate confluence of human activity. So what I saw might have been brick fill that helped build the highway. Or it could have been from any number of chemical spills. Or from paint. Or diesel oil dyed red to abide by government regulation. It could have been a trick of the imagination, a desire to crave something amongst the disgust of toxic petroleum refinement and the well-ordered monster of the shipping industry.

To eat dirt is also to make oneself container for these human additions to dirt, especially toxins that endure beyond the individual human lifespan. Clay works to eliminate toxins from the body because it has a large surface area and negatively charged ions. This is why you can dip wild potatoes in a clay sauce before biting in and expel the toxic alkaloids without being poisoned. Or why clay is used in the attempt to remediate oil spills, to cap landfills, or as a barrier to toxic seepage in New York City's waters when the Army Corps of Engineers buries contaminated dredged material the Corps pulls up to make way for larger and larger container ships coming in and out of New Jersey ports.[4] But once the residual of the anthropogenic, especially lead, creeps into and remains in the clayish object of geophagist desire, the appetite for ground becomes a hazard that lingers longer than the clay can.

Eating dirt in the midst of climate change brings the state of things directly into the body. This may be the case with all that is ingested, but with dirt, the human is especially vulnerable because the body becomes, like dirt itself, a container for industrial byproduct. The geophagist willingly consumes what is left over, not just from delicious weathering of old rock, but also from industrial products, like petroleum, that benefit from the slow geologic process of elemental transformation and then tend to leak back into the soil as poison. Eating dirt before the sedimentation of industrial material in the ground could have all kinds of deleterious effects on human health, but now, dirt exposes those who eat it to a myriad of heavy metals and chemicals such as lead, cadmium, and arsenic. These contaminants work upon bodies in varied ways as the climate changes. As temperature and rainfall fluctuate, so too does the pH and moisture of soil, affecting the bioavailability of both minerals and contaminants.[5] As climate change increases fluctuations in the environment, the dirt eater's body becomes measure of such inconsistency.[6] What that body reveals is an openness to toxicity that is inevitable regardless, but that hides behind the purity of the acceptable food items produced from the ground; food items that leave behind them chains of toxicity for other human and more-than-human bodies to deal with. Pesticides make clean-looking vegetables but, as we know, pesticides poison those organisms—human and not—who apply them, live near them, or drink water anywhere.

The dirt I saw in New Jersey is considered both abundant in contaminants and also largely infertile. Its color shows that there is not much organic matter present, just a lot of iron. This ground cannot be easily farmed, according to our agricultural practices; it must be fertilized to make up for its lack. But the human is also deficient in ways dirt can compensate. In much of the research on geophagy, anemia is listed as a root cause for the "disorder."[7] All that iron in red clay is supplement, appealing to iron-poor bodies like mine that turn to all the "wrong" things to make up for that lack: hiding the unspoken urge for a rusty nail or fork, choosing to chew on dandelion leaves or gnaw on bones and suck out the marrow to avoid more unacceptable forms of supplementation. If the desirous human lacks that for which dirt is the supplement, there are also ways in which this relationship is reversed. Wanting to meet teeth to this deprived dirt is perhaps a way for the calcium in those teeth to be a supplement to the lack of calcium in the soil—a desire to be needed by ground.

Sometimes geophagy helps people access minerals like iron, calcium, and zinc. Sometimes clay binds to important nutrients, making the body unable to digest them so that eating causes deprivation. Dirt-eating doesn't yield one particular result. Its undetermined effects are not dissimilar from those of consuming anything, but dirt-eating exaggerates the human organism's undetermined relationship to the milieu that becomes increasingly so as the effects of ecological catastrophe unfold. Food needs an elsewhere to dump its dirty surplus. Dirt is one of those grounds and the dirty practice of eating it helps redirect attention to that elsewhere that's also already metabolized. Eating dirt is but one form of intimate attention to what is deemed physically and conceptually background. Bringing that ground into the body, the geophagist does not necessarily welcome in the milieu more than anyone else, she just shows more agency in that permeation.

In the U.S. south, where topsoil had begun degrading very soon after colonialism began, slavery became one supplement for dirt-lack. Without glacial deposits of minerals, southern soils can't produce as well as other parts of the United States can. Alfred Cowdrey's environmental history of the southern United States explains that because of the lack in southern soil, the need for labor is greater—naturalizing institutionalized slavery.[8] He explains that "the chief peculiarity of southern soils, except the alluvial, is that they are old."[9] To account for both a peculiar and an old element of the south, he explains that because glaciers did not form in the region, most of its soils "lack the topping of minerals" that ice carried elsewhere.[10] In other words, one might say the south lacks a healthy supplement. Or, the peculiar, old soil of the south needs a supplement: the "peculiar institution" that is slavery of the Old South.

That institution of violence was also the context for pathologizing the practice of dirt eating. It was in the Americas that plantation owners began restricting people from eating dirt by covering their faces with metal masks.[11] These horrifying cages were promoted by the medical establishment. In 1836, *The London Medical Gazette* recommended attaching "a metallic mask or mouthpiece, secured by a lock" so as to create a "means of security for providing against their indulgence in dirt-eating."[12] That dirt eating should be considered an indulgence to be restricted by any means reveals the paranoia within the institution of slavery that the enslaved might find ways, even perhaps destructive ways, of exercising bodily autonomy.

Thomas Roughley warns in his 1823 instructional book, *The Jamaica Planter's Guide,* that geophagy "reduces the woman to a state of weakness, and barrenness, and makes her prone to idleness and disaffection to work."[13] For Roughley, dirt eating was the reason women refused to work or bear children. Perhaps the consumption of ground might also be understood not as a reason for, but as a means toward, resisting the conditions of slavery. Looking to historical examples where enslaved people were said to have eaten enough dirt to kill themselves, or to encourage their owners to sell them off, literary scholar Michelle Gadpaille argues that dirt eating within the institution of slavery was more than "pathology or taboo" but might "represent a means of negotiating power."[14] She writes that by taking possession of their innards if not their entire bodies, dirt eaters perhaps gained autonomy by bringing ground in through the mouth and foreclosing some of the ways their bodies were instrumentalized through work and reproduction. While I don't want to conflate a desperate response to the conditions of slavery with a means for rebellion, it seems significant that the geophagist perhaps made herself, to use Roughley's words, "barren and idle" by eating the land she was supposed to help make more productive and profitable.

Gadpaille reads the ways in which dirt eating also becomes medicalized when it endangers the purchase price of enslaved peoples. She writes that "plantation owners, denied a fresh supply of slaves by the trading ban of 1807, had an economic interest in exhibiting benevolence in their treatment of the remaining slave population" so that enslavers would attempt to heal instead of criminalize the practice.[15] The contemporary understanding of geophagy as a disorder is thus intimately tied to the system of American slavery and cannot be disarticulated from the ways in which land is forcibly toiled and spoiled, implicating individual bodies in that process.

Dirt's lack is always attended to through labor, but it is also supplemented later, beginning in the Gilded Age, by industrial fertilizers. Today, as the pressures of industrialization make dirt increasingly infertile, dirt is also, like the oceans, becoming more acidic. With increased global temperatures and rainfall, soils leach out their alkalines. When wetlands are drained for strip malls, agriculture, and highways, clays dry up and erode, removing the buffer between acid and earth.[16] Like the human body that, by ingesting clay, creates a wall between stomach lining and whatever else is ingested, the digesting Earth, at times, needs a clay lining. And in places like the Meadowlands—just north of where I saw the red dirt in New Jersey, that lining is being lost. Draining and toxic dumping have made the area a prime example of Superfund sadness. The poet James Schuyler notices those industrialized wetlands in his 1969 poem "Hudson Ferry." Instead of the grandiosity of the city and its waters that Walt Whitman or Hart Crane write about in their New York Harbor poems, Schuyler's poem commands an unidentified addressee to look at the New Jersey Meadowlands on fire "like a flushed cheek."[17] For Schuyler, a view of burning industrial space can still be tender as "a flushed cheek." He writes that nearby "a smokestack blows a dense dark blue," but he is less concerned with the smoke as polluter than with how to name the quality of its color or the way it moves "like hair / the bite-me kind springy or flung about."[18] The Meadowlands are depicted not just as a homogenous throwaway zone of capitalist refuse, but a specific place that requires precision. If not beautiful, the Meadowlands are particular, and that particularity, for Schuyler, can best be understood through insistent description. The poem reads as Schuyler's attempt to find out more about the Meadowlands beyond their status as industrial wasteland, perhaps similarly to how the geophagist comes to know a place through eating it. Humans digesting, like the gasping respiratory system of the poor wetlands of northern New Jersey, need a remediative layer, but such layers are also little bundles of toxicity.

Most geophagists don't eat topsoil, they dig holes down to the desired clay. But thinking about the dirt-eating human, I can't help but also think of the global crisis of topsoil loss. We all eat dirt, both literally, in the sense that it ends up in our food, and figuratively, as industrialized agriculture feverishly consumes fertile soil. Topsoil formation occurs in geologic time, that is, slowly; an inch forms every five hundred years. The United States is using up its reserve at ten times that rate, while China and India deplete theirs forty times faster than it can be replenished.[19] Topsoil loss occurs for a number of reasons. Sometimes it falls away, ending up in waterways and deposited elsewhere. Sometimes it is crushed into place by heavy machinery or asphalt. Other times it's contaminated by the addition of industrial materials or by salt or acid, or it loses too many minerals.[20] What does it mean to eat dirt in a time of massive soil loss? What does it mean to crave in this time of waning? Without answering what, exactly, is being craved—because it's different for everyone—one might say human organisms can be quite momentary containers for that which erodes at unprecedented rates. Or, we can be reminded, with a mouthful of anything good, that our bodies can be localized suspenders of the processes that outscale our temporal frame and waste away when those practices match human velocity.

The suspension of matter in the human body might also occur on a genetic level. Theories of epigenetics describe that which we consume, not just as matter moving through our bodies but as information that is able to regulate our genes. Sociologist Hannah Landecker, in her studies of scientific discourses around epigenetics, shows the ways in which the trend of this new thinking changes notions of the porousness of the organism. The theory holds that food, and whatever else is intentionally or unintentionally ingested, signals to our cells what the world outside our bodies is like, helping us to anticipate how to live in that environment.[21] Since this is thought to be especially true for developing fetuses, Landecker shows how epigenetics puts particular responsibility on pregnant bodies. The theory of "predictive adaptive response" speculates "that the maternal metabolic milieu educates the developing fetus as to the state of the world it will be born into," making the pregnant body responsible for feeding that fetus the proper information.[22] If, of course, those carrying fetuses around are charged with making sure the future is properly tended to, what kinds of information are being signaled to the geophagist's body, or to that of the fetus?

Landecker shows how eating is anticipatory because food helps the body live *"into* an environment in its future."[23] If eating becomes anticipation of a future milieu, how does dirt eating represent, and therefore start to posit into existence, some different world to step *into*; what sorts of futures are being invoked when dirt is consciously ingested? Perhaps one that gives primacy to a deep temporal frame, where the human body is seen as a porous conductor of toxicity and the ground as a necessary object of desire. In other words, there is, possibly, a productive sullying of the future in its digestive container of the present. New understandings of the anthropogenic entanglements with what used to be called nature mean not only that the human is denied purity from her environmental and social-historical milieu, but her genetics are dirtied with that milieu as well.

The contemporary poet, Elizabeth Alexander, explores this lack of purity from environment through a refusal to make abject the "dirty" practice of geophagy in her poem "Dirt-Eaters." The poem responds to a condescending newspaper article that reports a decline of geophagy among Black communities in the U.S. south. Alexander's poem doesn't simply reject the article, which was published in 1986 in the *New York Times,* but instead makes use of it. The poem quotes the article's claim that geophagy is a backward practice:

> "Most cultures
> have passed
> through
> a phase
> of earth-
> eating
> most pre
> valent today
> among

rural
Southern
black
women."[24]

In other words, southern Black women will catch up soon. Until then, we wait out the decline. Alexander's speaker resists this by looking back and responding with nostalgia. The poem responds to an outside intrusion of knowledge production about marginalized identities that get associated with an appetite for dirt. Instead of flat-out rejection of *New York Times* knowledge, Alexander mimics the columned form of the newspaper, digging down in her thin poem to a richer history and a richer earth where a connection to deep time can be pulled up and examined.

The form of "Dirt-Eaters" also mimics the task of the geophagist by digging into a neglected history and bringing it into the body of the poem, just as the geophagist brings deep time into her human body. Enjambing her lines, even her words, Alexander not only cracks into the narrative the *New York Times* aims to tell, refusing the knowledge of the "Ex / pert," but it also seems Alexander tries to crack into the practice itself.[25] As she digs down into geophagy, she uses short and sibilant lines that depict dirt eating as taking part in hushed networks of dirt and advice, networks populated by "gos / sips" and "(shoe-boxed dirt shipped north to kin)."[26] This dirt shipping is secret; it is safely nestled into Alexander's parentheses. But the secret isn't momentary, it stretches out like the hiss of "gos / sips" that are elongated through the line break. These secret and withdrawn lines of connection from person to person through clay are repeated in the gossips' rumors about the speaker's mother:

"Musta ate
chalk,
Musta ate
starch, cuz
why else
did her
babies
look
so white?"[27]

The repetition of the first two phrases makes the rumors feel repeated to the point of sedimentation, made into a tradition of their own: this is how things come to be known. And it seems to be one valid way to know, even if it's also kind of cruel.

Among the networks of people passing dirt, and ways of knowing about dirt, the speaker tries to forge her own lines of connection to geophagy. She says:

> Never ate
> dirt
> but I lay
> on Great-
> grandma's
> grave
> when I
> was small.[28]

Interestingly, she ties the memory of lying in the dirt to eating it. People lie on the ground all the time, especially as children, so it seems there is something particular about the grave. Perhaps it is because the earth is newly upturned, or because there is a recognition of significance beneath the top layer of soil. This child, laid out on the grave, mirrors her Great-grandmother's position. Such a mirroring has a temporal dimension, as the child, who directs attention toward the future, mirrors the grandmother of the past. If, as Lee Edelman argues, the Child is "the emblem of futurity's unquestioned value," what does it mean for that emblem to lie on a grave?[29] The backward-looking child, who also reminds us of the Child as container for the future, lying on the ground as her Great-grandmother decays below her, participates in the same kind of connection to expansive history as the dirt eater. Alexander points us to the grandiosity of the "Great" grandmother, whose name is still capitalized despite the fact that she is digested into small bits below the small child above. Those small bits are traveling, joining dirt across an expanse that is, like the speaker's grandmother, great. So it is in lying on the grave that the speaker brings her small, permeable body into contact with both the vastness of dirt and her own ancestry as a particular part of that vastness. This contact is made through the matriarch, reminding the reader that as Earth, so often deemed mother, digests the speaker's Great-grandmother, the figure of the dirt eater digests back, bringing the ancestral Mother into a present body. Even if she does not ingest it, this stanza seems to show the ways the speaker knows what it means to have dirt inside the body and to receive comfort from that fact.

The poem ends with the newspaper article again, quoting an older woman who is interviewed by the *Times*. The woman, Miss Fannie Glass, says:

> "I wish
> I had
> some dirt
> right now."[30]

In these lines, Alexander collects the deep social and temporal frame of digging for dirt to eat. The wish extends backward toward a practice that is being lost. The wish also gazes imaginatively toward a future that is as undetermined as the object of her desire. She simply wants "some dirt" and her lack of specificity speaks to the ways in which the desire for dirt might be a desire for an undetermined relationship between human and ground.

As Alexander helps show, the desire for dirt perhaps reflects a particular desire for intimacy to deep social and material histories as well as undetermined futures. The small appealing pile of dirt I saw in New Jersey is sedimented with countless histories that entangle the human organism with her dirty environment; and the ways in which we recognize this entanglement help shape the kinds of milieus we might show our bodies how to live into. If eating, as Landecker shows, is an anticipation of both our bodies and those of our offspring moving into future milieus, an appetite for dirt, in all the depth and indeterminacy it offers, perhaps opens up a future milieu that puts our current social-environmental conditions into relief against a deep temporal and spatial frame.

NOTES

1. See Claire Colebrook, *Death of the Posthuman: Essays on Extinction, Volume I* (Ann Arbor, Mich.: Open Humanities Press, 2014), 21.

2. As such, dirt eating is a willed invitation of the environment to move through the body. But such a relation already exists—whether or not it is willed.

Stacy Alaimo describes the human body as trans-corporeal, that is, in her words, "substantially and perpetually interconnected with the flows of substances and the agencies of environments." Stacy Alaimo, "States of Suspension: Trans-corporeality at Sea," *Interdisciplinary Studies in Literature and Environment* 19, no. 3 (2012): 476, doi:10.1093/isle/iss068 (accessed 14 August, 2016). Eating dirt is but one way of literalizing this relationship between organism and milieu.

3. *Diagnostic and Statistical Manual of Mental Disorders: DSM-IV-T*, 5th ed. (Washington, D.C.: American Psychiatric Association, 2013).

4. United States Army Corps of Engineers, "Historic Area Remediation Site (HARS)," (New York: Army Corps of Engineers New York District, n.d.).

5. Chiara Frazzoli, Guy Bertrand Pouokamb, Alberto Mantovani, and Orish Ebere Orisakwe, "Health Risks from Lost Awareness of Cultural Behaviours Rooted in Traditional Medicine: An Insight in Geophagy and Mineral Intake," *Science of the Total Environment* 566–567 (2016): 1468, doi: 10.1016/j.scitotenv.2016.06.028 (accessed October 12, 2016).

6. Linda Lorraine Nash, in her history of environment and disease, says that U.S. farmworkers of the 1950s who witnessed, and suffered the effects of, the industrialization of agriculture "read their bodies as a kind of instrument whose limits and illnesses measured the health of the land. Farmworkers located disease not in their own bodies or in their own communities but in a landscape that they found foreign and physically threatening." Linda Lorraine Nash, *Inescapable Ecologies: A History of Environment, Disease, and Knowledge* (Berkeley: University of California Press, 2001), 138. If the body is measure of the land's growing toxicity, this is perhaps especially true for the dirt eater. Like geologists using their mouths to identify soil and stones—wearing away at their teeth in the process—the dirt eater measures with her mouth. Taking ground into the body, containing it in the bloated stomach, the geophagist samples her milieu, coming to know its contents, toxic or otherwise, through the digestive system.

7. See, for example, Jo Hunter-Adams, "Interpreting Habits in a New Place: Migrants' Descriptions of Geophagia during Pregnancy," *Appetite* 105 (2016): 557–61, doi: 10.1016/j.appet.2016.06.033 (accessed October 12, 2016); Peter W. Abrahams and Julia A. Parsons, "Geophagy in the Tropics: A Literature Review," *The Geographical Journal* 162, no. 1 (1996): 63–72, doi:10.1023/A:1018477817217 (accessed 14 April, 2015); K. Kawai, E. Saathoff, G. Antelman, G. Msamanga, and W. W. Fawzi, "Geophagy (Soil-Eating) in Relation to Anemia and Helminth Infection among HIV-Infected Pregnant Women in Tanzania," *American Journal of Tropical Medicine and Hygiene* 80, no. 1 (2009): 36–43, doi: 19141837 (accessed 14 April, 2015); F. Tateo and V. Summa, "Element Mobility in Clays for Healing Use" *Applied Clay Science* 36 (2007): 64–76; Sera Young, *Craving Earth: Understanding Pica—The Urge to Eat Clay, Starch, Ice, and Chalk* (New York: Columbia Univerity Press, 2011).

8. Albert Cowdrey, *This Land, This South: An Environmental History* (1983; Lexington: The University Press of Kentucky, 1996), 58.

9. Cowdrey, 2.

10. Cowdrey, 2.

11. The artist Marianetta Porter recreated these muzzles out of silver and displayed them next to piles of dirt in her 2001 series, *Memory and Oblivion*. Patricia Yaeger briefly describes the show in the introduction to her book *Dirt and Desire: Reconstructing Southern Women's Writing, 1930–1990* (Chicago: University of Chicago Press, 2000), which begins with dirt eating (ix).

12. Young, *Craving Earth*, 73.

13. Thomas Roughley, *A Jamaica Planters Guide; or, A System for Planting and Managing a Sugar Estate, or Other Plantations in That Island, and throughout the British West Indies in General* (London: Longman, Hurst, Rees, Orme, and Brown, 1823), 118.

14. Michelle Gadpaille, "Eating Dirt, Being Dirt: Backgrounds to the Story of Slavery," *AAA: Arbeiten aus anglistic und amerkanistic* 39, no. 4 (2014): 3, https://www.jstor.org/stable/43025867 (accessed December 10, 2015).

15. Gadpaille, 6

16. A. Cook and K. Ljung, "Human Health and the State of the Pedosphere," in *Encyclopedia of Environmental Health*, ed. Nriagu, J. O. Elsevier (Amsterdam: Elsevier, 2011), 113.

17. James Schuyler, "Hudson Ferry," in *Freely Espousing* (New York: Doubleday, 1969), 49.

18. Schuyler, 49

19. Susan S. Lang, "'Slow, Insidious' Soil Erosion Threatens Human Health and Welfare as Well as the Environment, Cornell Study Asserts," *Cornell Chronicle*, March 20, 2006, http://www.news.cornell.edu/stories/2006/03/slow-insidious-soil-erosion-threatens-human-health-and-welfare (accessed May 3, 2015).

20. Cook and Ljung, "Human Health," 110.

21. Hannah Landecker, "Postindustrial Metabolism: Fat Knowledge," *Public Culture* 25, no. 3 (2013): 509–10, doi: 10.1215/08992363-2144625 (accessed December 14, 2015).

22. Landecker, 510.

23. Landecker, 515.

24. Elizabeth Alexander, "The Dirt-Eaters," *The Venus Hottentot* (Saint Paul, Minn.: Graywolf Press, 1990), 17.

25. Alexander, 19

26. Alexander, 18

27. Alexander, 19

28. Alexander, 19.

29. Lee Edelman, *No Future: Queer Theory and the Death Drive.* (Durham, N.C.: Duke University Press, 2004), 4.

30. Alexander, "The Dirt-Eaters," 19.

PLATE E3.1. Beatriz Cortez, *Cairn* inside Rafa Esparza's *Figure Ground: Beyond the White Field* (2017) at the 2017 Whitney Biennial. Courtesy of the artist.

ETUDE 3
Futurity Unknown

PLATE E3.2. Beatriz Cortez, still from *Childhood Bedroom / habitación de infancia* (foam board, wood, cloth, glue, paper, gesso, and digital video / cartón, madera, tela, goma, papel, gesso y video digital). Video installation. Courtesy of the artist.

PLATE E3.3. Beatriz Cortez and Rafa Esparza, *Nomad 13* (2017). Adobe bricks, steel, concrete, hammer, plastic, paper, soil, and plants: corn (maíz), black bean (frijol negro), prickly pear (nopal), sorghum (maicillo), amaranth (amaranto), quinoa, chayote squash (huisquil), chia (chía), chili pepper (chile), yerba buena, yerba santa, sage (salvia), and silk floss tree (ceiba) (104 x 84 x 96 inches). Courtesy of the artist.

PLATE E3.4. Beatriz Cortez, *Memory Insertion Capsule* (2017). Steel, archival materials on video loop. Courtesy of the artist.

CHAPTER 14

The Memory of Plants
Genetics, Migration, and the Construction of the Future
BEATRIZ CORTEZ

> The little green limbs that populate the planet and
> capture the energy of the Sun are the cosmic connective
> tissue that has allowed, for millions of years, the most
> disparate lives to cross paths and mix without melting
> reciprocally, one into the other.
>
> —Emanuele Coccia, *The Life of Plants:*
> *A Metaphysics of Mixture*

For many years now, nomadism has been an important concept in my thinking. I arrived there via queer theory. Queer theory opened up the possibility of exploring a space of blurriness and nondefinition that fit the experience of the migrant. Being a nomad was conceptually akin to the refusal to exist within the given diagrams of compulsive heterosexuality. Movement has been a central idea in my thinking, one that unites my theoretical framework and my life experiences. As a result, my reflections about memory distance themselves from the institutional concept of historic memory. I approach memory as a concept that stretches temporalities from the past toward the future, carries polyphonies, multiple textualities and languages, and different materials and states. When imagining memory as something that is in constant flow, it becomes nomadic, it unfolds the past but also opens up to posthuman possibilities projected beyond the Anthropocene, toward a future without humans.

As an artist, my work explores the memory of growing up during the civil war in El Salvador, and my experiences of migration, first to Arizona, where I lived for almost eleven years before coming to Los Angeles in 2000 via Detroit. But my work is also about the experience of simultaneity: being in two different timeframes, living in two places at once, San Salvador and Los Angeles, existing in two or more—but different—versions of modernity, in different cultural worldviews, moving back and forth within different technologies. And more importantly, my work is about the future: I build memory so I can imagine possible futures.

Three concepts are central to my work: time, simultaneity, and movement, and I am interested in how issues such as race, class, migration, and gender cut through those concepts. The environment has been an important source of metaphor for me. Even when I work with industrial steel, it is with the intent of returning this industrial material that has been artificially extracted from the earth back to the land. Building rocks with steel is an example of that gesture. However, it is through my work with plants that I have understood that our mark on this planet is not only written on the sedimentation layers or on the rock but also on the living environment. After all, a garden is an artificial arrangement of plants, a form of human intervention with cultural and historical content. Plants have allowed me to build metaphors about eugenics, about labor, migration, hybridity, ancient knowledge, and more, as I will discuss in the pages to follow.

In order to move through space and time, and to explore simultaneities, the ability to intervene in the chronological order of time is central to my work. One of the philosophers who questions the way modernity imposes a chronology and a universal narrative of time is Peter Osborne. He argues that the effort to establish a narrative about modernity toward progress, toward development, is also a colonial and homogenizing effort to erase diversity.[1] Rather than a chronology, Osborne is interested in exploring different modernities that coexist simultaneously. Ideas about speculative realism, the posthuman, philosophies of extinction, invited me to question the imposition of humanism, liberalism, or the Enlightenment as the only way, a colonial way, to understand reason, and with it, to understand time. Instead, I became interested in other ways of understanding time, circular conceptual constructions of time, multiplicity, and simultaneity.

Death as a way of becoming and extinction as an imaginary began to populate my mind as I read about the nonhuman and about a future after humans, particularly through the work of Claire Colebrook. I accepted her invitation to imagine a nonhuman geologist of the future reading our mark on the planet. This nonhuman geologist of the future would read the sedimentation layers, the rocks, and the stones for traces of the Anthropocene, for traces of human life on the planet.[2] But its job will not be easy, I imagined, and lava came to my mind as a material that would confuse this geologist of the future and would blur the records, but more importantly, would break with the chronological order of time. The lava from the past spit out by a volcano (there are always volcanoes in the landscapes that I imagine) would change the order of the layers that write the temporal history of our passing through the earth. And this is how I began to explore different ways in which matter breaks chronology.

I built *Cairn* first outside the Brand Library in Glendale, California, and later inside Rafa Esparza's *Figure Ground: Beyond the White Field* at the 2017 Whitney Biennial. It was a mound of stones serving as a landmark. Indigenous peoples have built cairns for centuries to mark the way on a trail, to mark sacred or burial ground, or as other forms of landmarks. Within modernity, a cairn is often used to delineate private property or to trace the border between two nations. At the turn of the twentieth century in Los Angeles, this Indigenous way of building with river rock was often used to build Craftsman and Spanish Colonial Revival houses in what came to be known as a vernacular style of construction. It was an ancient form of construction applied to a modern architectural style. It was a syncretic way of building within and outside of modernity in a simultaneous way.

As I thought about materials, plants became an important medium for me, and part of my language. They were the symbol of the tropics and of the passage of time in my video installation *Childhood Bedroom*. And when I built *Unknown*, a sculpture with soil and seeds that was meant to honor the dead who have been thrown out like trash during war and migration, plants transformed the sculpture into a garden that flourished in honor of someone who was not there, symbolically crossing the barriers of time and space.

In her book *The Death of the Posthuman: Essays on Extinction*, Australian philosopher Claire Colebrook argues that human beings are the authors of our own extinction (due to our desire to eat fats and sugars, and to our overconsumption) and also of the destruction of the environment that we need in order to exist.[3] For Colebrook, it is important to think of the possibility that humans are about to disappear, and that the human era—the Anthropocene—is about to end. Nevertheless, she reminds us that human beings are not necessary for the planet's existence or for the world to continue: other eras without us have preceded us and other eras without us will follow.[4] In terms of the time that the planet has existed, we have been but a brief lapse in its history. It is from this perspective that I am interested in thinking that humans are not necessary for memory to exist, that it is possible to think of a posthuman memory, or a nonhuman memory: for instance, the memory contained in lava or the memory of plants.

As she imagines a nonhuman world, Colebrook's reflections center on vision through the human eye. She defines the eye, conceptually, as the organ that organizes the world, a synthesizer that reads, theorizes, and systematizes everything it sees.[5] And she says that "the human animal or human eye is torn between spectacle (or captivation by the mere present) and speculation (ranging beyond the present at the cost of its own life)."[6] On the one hand, this eye that synthesizes and digests the world for us is corporeal, and through seeing reproduces our humanity. Her perspective on the human eye shows what it sees but also the human body that contains it, much in accordance with the way in which Judith Butler argues that a photographer, even if not visible within the frame of a photograph, is always present in the image.[7] Thus, vision runs the risk of reconstituting humanism. In her conversation with Bergson, Colebrook reminds us that "the human eye organizes the world into conceptualized units, mastering the world by reducing difference."[8] The eye takes what it sees, pays attention to some details, puts some images into focus, and organizes and digests what it sees into a world that is coherent and unified, a world without differences, a human world. From this perspective, the eye that sees is the one that constitutes the human subject.[9]

But on the other hand, Colebrook says, another way to deal with the eye is to think of it as a machine. Engaging the work of French philosopher Gilles Deleuze, she wonders if it is possible for nonhuman perception to exist, if it is possible to imagine a world without us, not the world as our environment or surroundings but in an era where human beings no longer exist or, as she states, a world that would "open us up to the inhuman and the superhuman durations,

[. . .] beyond the human condition."[10] In this sense, the nonhuman eye, the machine eye, could edit in more difference, could potentially see other details and other diversities that the human eye does not see, it could complicate things and see beyond human limits.[11] For instance, a *selfie* would be a perspective par excellence within the diagrams of humanism. By contrast, the image generated by the X rays that our dentist takes, or the landscape projected by the camera during a colonoscopy procedure, exemplify the vision of a machine eye.

And so, Colebrook imagines a machine eye that exists when humans have ceased to exist on the planet. It is an eye that reads the sediments that humans have left through their time on the planet, the remains of the Anthropocene, what she calls the scars of the strata of the earth that mark human life. This machine eye would move through the earth as a researcher moves today through an archive, as an archaeologist moves through a site, and would read the world, its anthropologic scars, its survival in spite of human existence on its surface. Colebrook imagines this machine eye as a geologist of the future.[12] This nonhuman geologist of the future would detect other rhythms, would take different points of view about what has been recorded on the earth. Maybe this geologist of the future will be like one of those rovers that move through Mars, one that will read our present at a moment when it will have become part of the past. It will also read what has not yet been recorded at a time when it will have become the past. This world will not be seen from a human body, it will not be "a world for a body,"[13] it will be an impersonal image. She invites us to think about extinction and the possibility of abandoning subjectivity as we have understood it through the lens of humanism.

In *Theory and the Disappearing Future,* Colebrook addresses deconstruction, not from a tradition linked with Derrida, which after all reconstitutes subjectivity and context, and by doing so, reinstitutionalizes the diagram of identity practices inscribed on the subject within the humanist philosophical tradition. That is, Derrida's deconstruction moves toward the past. On the contrary, Colebrook is interested in understanding deconstruction as future oriented, from a perspective that is closer to the teachings of Paul de Man and his insistence that we approach culture, images, texts, and landscape "as *if* it detached from all humanity, as if—to use Deleuze and Guattari's terms—it was without a people, as though the population were still missing."[14] In other words, to approach everything as if the humans that were meant to perceive it had not yet been born. That is, to approach everything as if it were part of the future, especially, a nonhuman future that would allow us to imagine a different, new form of vision. Rather than restoring everything to its original intention, which would be equal to assigning it a fixed identity, this would give new life to a text or an image each time we approach it.[15] It would open possibilities for becoming when facing each text or work of art, always toward the future. It is with this process in mind that I would like to approach the knowledge and the wisdom embedded within plants.

The work of Brazilian artist Maria Thereza Alves has impacted my reflections about the memory of plants, the posthuman life of plants, and the nonhuman possibilities held in plants. Alves' project titled *The Seeds of Change,* installed at ports in different European cities, including Bristol, Marseille, and Liverpool, invites the viewer to reflect upon the legacy of colonialism through the displacement of plants through their seeds traveling as part of ballasts, or disposable materials that were discarded by the ships upon their return from the Americas and other locations to where they took humans to be sold and enslaved. Life in the dormant seeds remained latent for centuries, and the artist-turned-gardener was able to germinate them once again, in order to create beautiful gardens displaced from their sociopolitical histories, ancient Indigenous plants growing in European port towns, a visible remnant of slavery.

It was in part due to these reflections about Maria Thereza Alves's work that I began to think about the memory of plants in the context of Central American migration, about how the plants preserved the memory of a migratory experience that followed a very similar trajectory to my own.

After the publication of the work *The Life of Plants: A Metaphysics of Mixture* by Emanuele Coccia, I began to think about plants as the lungs through which an atmosphere emerged, what he called "the essence of cosmic fluidity, the deepest face of our world, the one that reveals it as the infinite mixture of all things, present, past, and future."[16] From his perspective, to breathe is "to embrace in one's own breath all the matter of the world."[17] For Coccia, plant leaves are "a climatic laboratory par excellence, the oven that produces and frees it into space, the element that renders possible the life, the presence, and the mixture of an infinite variety of subjects, bodies, histories, and worldly beings."[18] Reading Coccia I began to think about plants and their cosmic mission of pushing further the becoming for the present, past, and future of the planet. For me this confirmed the porosity of all bodies, and reminded me that it would not only be the stones, the sedimentation layers on the planet, as Colebrook had invited me to imagine, but also plants that would preserve the memory of the Anthropocene, that would circulate that memory, its physicality and fluidity, through a becoming that would extend beyond the human era.

But it was also in the archives that the history of migrating plants impacted me profoundly, especially when I realized that it was linked to colonialism and to a deep racism founded on white supremacy. A particular last name called my attention in the archives: Popenoe. I had been to the Casa Popenoe Museum in Antigua, Guatemala several times. It was the house of Wilson Popenoe, the director for more than twenty-five years of the Panamerican School of Agriculture, also known as El Zamorano, established by the United Fruit Company in Zamora, Honduras. It was in these archives in Los Angeles that I realized that the Popenoe last name not only evoked the Zamorano or the United Fruit Company but also a history of racism and eugenics. Wilson's brother Paul Popenoe had served as secretary of the Human Betterment Foundation during the 1930s in Pasadena, California. It was an organization that advocated for the forced sterilization of Blacks, Latinos, and Indigenous peoples, as well as migrants, the ill, and the poor. It also evoked a history of colonialism, since both Paul and Wilson Popenoe had participated in a program of Agricultural Explorers that was fostered by the United States.

The Department of Agricultural Explorers was created by the United States Ministry of Agriculture in 1898. Each of these explorers was charged with the task of travelling the world in search of new crops that could be transplanted to the United States. At the time, there was a man by the name of Fred Popenoe who had a nursery in Altadena, a neighborhood in the northeast of Los Angeles. In

1911, encouraged by exploratory fever and demand, Fred sent two of his sons, Wilson and Paul Popenoe in search of dates to grow in California's Coachella Valley. They traveled through Iraq, the Persian Gulf, the Red Sea, and northern Africa. The result of their first expedition was the transplantation of date crops to California, creating an oasis with palms in an inhospitable desert. However, they also established there, in places such as Coachella, La Quinta, and Palm Desert, among others, an orientalist culture and festival that included belly dancers and camel rides. The Popenoe brothers, embodying the symbolic link between the imperialistic extraction of natural goods in Central America and the rejection of Central American immigrants in the United States, also advocated that genetic experimentation would allow them to build a world for whites, whom they considered superior.

Afterward, Wilson Popenoe dedicated himself to genetic experimentation with bananas in Central America, and to the transport of plants to California, like the avocados that nowadays are known as California avocados. Meanwhile, Paul dedicated himself to experimenting with humans, contributing to the sterilization of a great number of people in California. Their work, through the Human Betterment Foundation, sparked great interest in Germany during the 1930s and was allowed to continue in California under the protection of the law until the late 1970s. However, when World War II ended, bringing with it eugenics' decline in popularity, Paul Popenoe devoted himself to expanding his eugenics project to phase 2, by establishing the first marriage-counseling programs meant to promote the wedding and procreation of superior couples. He created the National Family Defense Fund in Hollywood, and began publishing a series of columns in the newspaper titled "Can This Marriage Be Saved?" as well as radio and reality TV shows titled "Divorce Court."

These ideas about the memory of plants have nourished my art practice. I created *Nomad 13*, a collaboration with Los Angeles–based artist Rafa Esparza, a sculpture made in the shape of a space capsule, a cosmic garden that preserved the knowledge and culture of ancient Indigenous peoples in the Americas for the survival and pleasure of the humans of the future. I also created *The Memory Insertion Capsule,* a steel structure that takes on the form of a space capsule but also brings together Indigenous forms of architecture, Spanish

Colonial design, Craftsman architecture, industrialization, and precarious tent living, evoking in this way multicultural coexistence in the City of Los Angeles, as well as the refugee and housing crises. It also includes a visor that simultaneously evokes the Mayan glyph for zero and a machine eye, one that allows the viewer to see fragments of this historic relationship between the United States and Central America, as well as the role of the Popenoe brothers in this colonial and imperialistic enterprise. By watching the video, viewers are gifted memories taken directly from local archives about immigration, racism, scientific experimentation, agricultural exploration, the history of the United Fruit Company, marriage counseling, and eugenics.

From the perspective of the United Fruit Company, as well as Western cultural and scientific legacies, plants are discussed as beings with fixed identities, with borders and national identities. They are seen as impermeable bodies, completely isolated from the outside world. From this perspective plants are classified by their place of origin, grown for diverse purposes that include generating profit within capitalism, as is the case with monocrop agriculture, or to denote wealth and power, as is the case with large gardens in front of mansions, with bowling greens and geometric designs. However, there are other reasons to have gardens, not all inscribed within capitalism: for nostalgia of a landscape that only exists elsewhere, for pleasure, for food production, for cultural reasons, for generosity. I am interested in gardens, not only for how they index the past, or the origins of each plant, but also the future. These are gardens that take shape for the humans of the future. As they move through the Anthropocene, they preserve the memory of the humans that lived on the planet for the machine eye of a future geologist or for the atmosphere that beings of the future will breathe.

In this way, gardens are intermediaries between the Anthropocene and the mark that humans leave on the planet. Plants are survivors of repression, prohibition, colonization, slavery, capitalism, imperialism, eugenics, and genetic experimentation. They have survived, reminding us of the porosity of their bodies and the futility of borders, as Karen Barad explores in her work about the queer nature of atoms,[19] or as Coccia argues when writing about the mixture of matter in the atmosphere.[20] And so, plants leave evidence of our migrations as the natural condition of the planet. And I would like to think that in our absence, they will remain latent, preserving future life and the memory of colonialism, racism, and migration, but also the memory of our generosity and our courage to survive.

NOTES

1. Peter Osborne, *The Politics of Time: Modernity and Avant-Garde* (London: Verso, 1995), 16–17.

2. Claire Colebrook, *Death of the Posthuman: Essays on Extinction* (Ann Arbor, Mich.: Open Humanities Press, 2014), 24–28.

3. Colebrook, 21.

4. Colebrook, 23–24.

5. Colebrook, 13–16.

6. Colebrook, 14.

7. Judith Butler, "Photography, War, Outrage," *PMLA* 120, no. 3 (2005): 822–27.

8. Colebrook, *Death of the Posthuman,* 16.

9. Colebrook, 21.

10. Colebrook, 23.

11. Colebrook, 23.

12. Colebrook, 24.

13. Colebrook, 27.

14. Claire Colebrook, "Introduction," in *Theory and the Disappearing Future: On de Man, On Benjamin,* by Tom Cohen, Claire Colebrook, and J. Hillis Miller (London: Routledge, 2012), 8.

15. Colebrook, 8.

16. Emanuele Coccia, *The Life of Plants: A Metaphysics of Mixture* (Cambridge: Polity, 2019), 27.

17. Coccia, 54.

18. Coccia, 28.

19. Karen Barad, "Nature's Queer Performativity," *Kvinder, KØN & Forskning* 1–2 (2012): 25–53.

20. Coccia, *The Life of Plants,* 54.

Coda

CAROLYN FORNOFF, PATRICIA EUNJI KIM, AND BETHANY WIGGIN

We must realize that the world as it is isn't worth saving;
it must be made over.

—John Rice

In attempting to describe the predicament we are in,
therefore, the best we can do is to speak in the future past
(*futur antérieur*): "it will have been the best of times, and
the worst of times, but it will have been our time, though
we were only passing through."

—Rosi Braidotti

It will have been a time of our *great derangement*, a time of collective madness.[1] It will have been a time of working on problems long in the making but also urgently compelling responses, some of which will necessarily be slow.[2] It will have been a time requiring us to think in the plural: with codas and queer returns, in multiple time signatures, within and across an array of timescales. It will have been a time in which truth will sometimes have been stranger than fiction. It will have been a time in which science and speculative fictions flourished. It will have been a time when experimental cultural forms were cultivated, in academic registers too, as antidotes to realist genres' recalcitrant representations of the ongoing ecological crisis.[3]

It will also have been a time in which experimental research collaborations blossomed, mutated, and transformed. As posthuman philosopher Rosi Braidotti—from whom we have borrowed this coda's second epigraph and thought with in our introduction—writes, "'We' need to connect critique to creativity and invent new ways of thinking, inter-woven lines of posthuman critique, because we cannot solve problems in the same language we used to create them."[4]

We, this coda's coauthors and *Timescales* coeditors, are not, and will have not been, in this alone. As our web of citations throughout this collaborative volume signals, we have been, are, and will continue to be inspired by feminist environmental scholars and activists. *Timescales* will have been the direct result of the generosity emblazoning the banner of the best recent work in environmental humanities. It will also have been the result of our shared belief—as Ursula Heise describes in her survey of the environmental humanities, drawing in turn on the collaborative statement by Deborah Bird Rose, Thom van Dooren, Matthew Chrulew, Stuart Cooke, Matthew Kearnes, and Emily O'Gorman—that we must make "an effort to inhabit a difficult space of simultaneous critique and action."[5] Our collective sense of action has been most directly inspired by the authors who have contributed their own experimental critiques and/as actions to *Timescales*. Together, we've explored what collaborations between the sciences, arts, and humanities can look like in the environmental humanities. Several models are included within this volume: a geoscience perspective articulated in a register accessible to readers beyond the discipline (Dmochowski and Evans), a humanistic critique of technoscience's imperative to maximize yield (Telesca), and, perhaps most radically, a coauthored chapter by an oceanographer and a literary scholar who advocate for interdisciplinary pessimism, the abandonment of solution-oriented collaboration, and the use of exploratory chitchats (Bell and Pavia).

At different scales of time and of place, all of our volume's authors grapple with research subjects and objects implicated in the changing climate and differently scaled, interwoven phenomena. These models scale up and down and attempt to think simultaneously with multiple timescales that diverge from the tradition of linear historicization. *Timescales* thus presents a sustained experiment in bringing multiple, overlapping, imbricated, and entangled timescales together and placing them in conversation. No less than the individual essays, the structure of the volume as a whole advances this model of intellectual experimentation, gathering different disciplines—with their supposedly discreet timescales, methods, and objects—into conversation and productive friction.[6] On an individual level, many authors tackle the simultaneities and incommensurabilities of different timescales head-on. Collectively, they share a sense of the urgent need to make times touch, to queer the tidy progression of timescales marching in neat succession.[7] As some of our contributors have shown, toggling between and across different scales is essential to living in a time of environmental crisis, as well as to imagining what the future might hold. The challenge that we now face is to learn, as Dipesh Chakrabarty put it, to think "disjunctively."[8] Ömür Harmanşah embraces disjunctive timescales in the genre of history writing, bridging together the expanse of the past, present, and future through the concept of temporal percolation, or "deep time leaking into the present through ongoing material entanglements and spatial proximities." Jane Dmochowski and Dave Evans acknowledge the gaps within deep, geological histories, which nevertheless signal the urgent need to take climate action within a much shorter span of time.

Speculative projections of a future marked by discontinuous environmental conditions—rising waters, extreme weather events, mass extinction—in many cases require aesthetic techniques that stretch beyond anthropocentric perspectives. Experimental play with zoomed-out viewpoints is by no means unique to the twenty-first century, but has been a keystone for thinking about humanity's past and future planetary place. Charles Tung elaborates that for the modernists of the early twentieth century, the "far-futural scope" of the time machine provided one such aesthetic tool for projecting beyond the narrow lens of human history. Likewise, Beatriz Cortez's speculative sculptural practice imagines how a nonhuman geologist of the future might be perplexed by the misleading temporal clues left behind by lava that blurs and "breaks the chronological order of time." Cortez's projections of Indigenous technologies into the far future echoes Déborah Danowksi and Eduardo Viveiros de Castro's observation that if we are looking for models for future survival under calamitous conditions, we should look to Amerindians, whose world ended, yet they carried on "in another world." That is, they witnessed the apocalypse of colonization and have learned lessons of survival that can "serve us as example and warning regarding contemporary processes."[9] These insights highlight the centrality of postcolonial and critical race theories in current conceptualizations of climate change; as we now know, the genocide of 56 million Amerindians by 1600 caused the planet to cool, evidencing the human role in climate change prior to the Industrial Revolution.[10]

Moreover, the timescales and lifeways of nonhuman animals have drastically altered in a relatively quick span of time during the Anthropocene. It is only by noticing the absence of nonhuman life-forms, and in some cases listening for it, as Wai Chee Dimock argues, that we might begin to mourn the irreversible loss driven by human practices of consumption and habitat destruction. Dimock proposes that extinction sounds different—it is quiet; it is a withdrawal of what once was. Indeed scalar toggling demands an attunement to the world that engages multiple senses and perspectives.

For others, thinking about the far-off future does not require a view of the planet from a nonhuman eye and instead centers human trans-corporeal relationships with environments. Following Stacy Alaimo's work, transcorporeality defines the ways in which human bodies are enmeshed with nonhuman others and landscapes in a way that re-orients subjectivity.[11] Contemporary hoopwalkers of the American West, Paul Mitchell tells us, see their own movements on the hoop "as work on the self in transition [. . .] implicating transitions of the individual, social, generational, and ecological into each other." Although such queer environmental and activist practices of re-wilding problematically (re)perform Native American practices, such lifeways represent an Anthropocene chronotope, or entanglement of human and landscape. The deep past contained in the sediments of the earth, as well as the toxic residues of human practices of moder-nity, is an archive that Iemanjá Brown ingests. Geophagia and bodily cravings for the earth enact an explicit recognition that the human body is composed of and through the nonhuman, engagements that are often racialized and gendered.

Ecological change has likewise provoked imaginative experiments in living otherwise, as artists offer speculative meditations across media about the stakes of cross-disciplinary action. The aesthetic techniques shared throughout the volume have translated the large geographical, temporal scales of the Anthropocene into terms that are made perceptible at an individual human scale through empathy and intimacy. *A Period of Animate Existence,* a theatrical performance that gathers force from the confrontation between young and old bodies sharing the stage, elicits both grief and laughter from viewers. Humor, the uncanny, and irony draw out a set of ethical questions about taking action in a changing climate. In a similar vein, Mary Mattingly's *WetLand* was a floating experiment and utopian vision of living that sank into the waters of the Schuylkill River. These practices represent hope for a future perfect, as artists encourage viewers to embrace the open-ended temporalities of speculative experiments, including their possible failures.

Climate change has shaken up our temporal orientation, compelling us to think about divergent futures while confronting animate archives of the deep past, sometimes literally, as thawing permafrost releases so-called zombie pathogens, vivifying hundred- or thousand-year-old bacteria. The scalar collision of these timescales requires us to zoom out or to straddle multiple temporalities. Attuning to time in the plural necessitates experimental alliances among scholars, artists, and activists. The timescales of these collaborative experimentations in living or thinking otherwise are open ended, and their outcomes uncertain. We must be willing to let them sink, fail, or awkwardly peter out. This is the task at hand: to continue the daily work of maintenance but also to play, tinker, build, grasp at straws, weave threads together, or, as Donna Haraway urges, "stay with the trouble."[12]

Chitchat Codetta

CAROLYN FORNOFF: As I sit here at my desk in Illinois, I can see wisps of green poking through the ground, a welcome sight that winter might soon be over. Friends in Houston are sharing photos on Facebook of a toxic gas cloud looming over the city, likening it to the "airborne toxic event" that appeared in Don DeLillo's *White Noise*. An energy lobbyist, Andrew Wheeler, has just been confirmed as the chief of the EPA. It's my second time teaching a novel about whaling and extinction, *Mundo del fin del mundo [World at the World's End]*, written by Chilean novelist Luis Sepúlveda in the late eighties. Thirty years later, a student pulls up an article on her smart phone of a beached beaked whale that washed ashore in the Philippines with eighty pounds of plastic in its stomach. Literature, the classroom, and the internet make these far-flung times and spaces—Chile and the Philippines, the eighties, the lifespans of whales and plastic, the rare visibility of toxicity—more palpable here in the Illinois spring. Soon it will be time to plant seeds. I'm new to this part of the country, so it may take years to figure out what will take in my backyard. And that might change as things become warmer and wetter.

BETHANY WIGGIN: Stengers says we need to be like gardeners in cultivating slow science. I like how she, and you, are suggesting we return to the garden—but other gardens in other worlds—even as climate zones are migrating. (Did you know good pinot noir now grows in Washington? Probably north of Vancouver, too, by now.) You both remind me of Candide, who was quite sure that we don't live in the best of all worlds and told his friends *"il faut cultiver notre jardin."*

Sometimes I teach a class on censorship and the book, and *Candide* might be my favorite book on the course list I made. (Sadly censorship feels all too topical these days—just this morning, I reposted on social media a stunning report published in the *New York Times* that also drew on work by the terrific team at the Center for Biological Diversity about the ongoing suppression of science at federal agencies, this time of Fish and Wildlife scientists' reports on pesticides and endangered species.) Candide has been exploring the "new world" and witnessing the brutal violence that was the everyday world of Indigenous and African enslaved peoples made to extract silver and cultivate sugar. Your gardening reminds us we're still in that moment too, *in the wake,* as Christina Sharpe writes. We're still in the tow of that history that isn't over. So much for progress.

I just stepped away after teaching a class on the Schuylkill River and environmental history with twelve teachers who work in the School District of Philadelphia. We had read Cronon's classic "The Trouble with Wilderness," and we kept returning tonight to this question of how we might cultivate otherwise. Can we learn to live with and love not only the land? What about our waste sinks and inhabited seas?

CF: In music, the coda usually recapitulates the exposition, echoing its themes. It makes me think about where we began: together, in Philadelphia, perturbed by the newly elected Trump administration. It was difficult to feel optimistic in that moment. But we were working to build something new. *WetLand* had yet to sink, PPEH was not yet fully afloat. It felt like a moment of impasse, as Lauren Berlant describes it in *Cruel Optimism*, a sort of improvisational treading of water. Response to the present, in writing, in politics, in teaching, takes time.

PATRICIA EUNJI KIM: I'm late to this conversation. I've been late to a lot of things these days—meetings, doctor appointments, my own deadlines. In my defense, there's too much to get done! Here's my list: do my laundry, read the news, post on social media, return library books, call my mother, defend my dissertation, exercise, water my plants. My plants are dying. A few years ago, the plants in my parents' garden were dying, too. I went back home to southern California and everything was desiccated. The park where my dad taught me how to ride a bike had turned grey. I know, I *know*: *oh no! The suburban lawn is dying!* But it was still hard for me because it changed so quickly and no one warned me. I wasn't ready.

Last year, I visited Seoul for the first time in almost a decade. It was hot and the air wasn't great. But it felt wonderful and weird to wander around the city where my parents grew up. I imagined my parents as teenagers, walking up and down Seoul's many small hills and winding alleyways (without the air pollution). My grandmother would sing this song over and over again: "내 님은 누구일까? 어디 계실까? 무엇을 하는 님일까? 만나보고 싶네. 신문을 보실까? 그림을 그리실까? 호반의 벤치로 가봐야겠네." It's one of those cheesy love songs in which the singer asks where and who her next lover will be. She has no idea. But then she says something like, "well, I better go wait on the lakeside bench." Why she'd want to do that, I don't know. But neither does she. She's just waiting for something to happen while she sits near a body of water. Pretty soon, we'll all be wading through rising waters.

CF: It is hard not to feel impatient. And yet here we are, carrying on, brought together by this Google Doc and these questions that stretch into the far future. While it feels like we are at the end of this project, we are actually lodged in the midst of these timescales of concerns. But together.

BW: I just want to say thank you. I miss you guys. I'd do it all over again. But only with collaborators like you :)

NOTES

1. Amitav Ghosh, *The Great Derangement: Climate Change and the Unthinkable* (Chicago: Chicago University Press, 2016).

2. In her "plea for slow science," Isabelle Stengers invites us to explore how research and knowledge production might "reclaim scientific research" from the compulsions of fast science. Her exposition of "activists engaged in reclaiming operations" resonates with our aims in *Timescales*. She explores the generation of care and concern, how gathered individuals come to "allow the issue at the heart of their meeting the power to matter, the power to connect everyone present." Slow science refuses to take such moments superficially or deem them irrational and so "unfit for knowledge." Like Stengers's slow science, we are interested not in learning how to define these moments "but rather how to foster them." Isabelle Stengers, *Another Science Is Possible: A Manifesto for Slow Science*, trans. Stephen Muecke (Cambridge, U.K.: Polity Press, 2018), 123.

3. Ghosh considers how, with the rise of realist literary forms, the nonhuman was relegated from "the mansion of serious fiction" and relegated to "the outhouses to which science fiction and fantasy had been banished." In the face of the scalar challenges posed by the Anthropocene, the realist novel is unable to scope "forces of unthinkable magnitude that create unbearably intimate connections over vast gaps in time and space" (*The Great Derangement*, 66, 63). We note, as Heise emphasizes, that the mansion of serious fiction has many rooms that exceed the dimensions of literary realism. Ursula Heise, "Climate Stories: Review of Amitav Ghosh's 'The Great Derangement.'" *b2o*, February 19, 2018, https://www.boundary2.org/2018/02/ursula-k-heise -climate-stories-review-of-amitav-ghoshs-the-great-derangement/ (accessed March 27, 2019).

4. Rosi Braidotti, "Posthuman, All Too Human," 2017 Tanner Lectures, 21, https://tannerlectures.utah.edu/Manuscript%20for%20Tanners%20Foundation %20Final%20Oct%201.pdf (accessed February 4, 2019).

5. Ursula K. Heise, Introduction to *The Routledge Companion to the Environmental Humanities Routledge*, https://www.routledgehandbooks.com/doi/10.4324 /9781315766355.ch101 (accessed October 30, 2018).

6. Anna Lowenhaupt Tsing, *Friction: An Ethnography of Connection* (Princeton, N.J.: Princeton University Press, 2004).

7. "Touching time" borrows from Carolyn Dinshaw's seminal patchwork of past and present, *Getting Medieval: Sexualities and Communities, Pre- and Postmodern* (Durham, N.C.: Duke University Press, 1999).

8. Dipesh Chakrabarty, "Postcolonial Studies and the Challenge of Climate Change," *New Literary History* 43, no. 1 (Winter 2012): 2.

9. Déborah Danowski and Eduardo Viveiros de Castro, *The Ends of the World*, trans. Rodrigo Nunes (Cambridge, U.K.: Polity Press, 2017), 106.

10. Alexander Koch, Chris Brierley, Mark M. Maslin, and Simon L. Lewis,

"Earth System Impacts of the European Arrival and Great Dying in the Americas after 1492," *Quaternary Science Reviews* 207 (March 2019): 13–36.

11. Stacy Alaimo, *Bodily Natures: Science, Environment, and the Material Self* (Bloomington: Indiana University Press, 2010).

12. Donna J. Haraway, *Staying with the Trouble* (Durham, N.C.: Duke University Press, 2016).

CONTRIBUTORS

JASON BELL is a student at Harvard Law School. He received his PhD in English from Yale University.

IEMANJÁ BROWN is a visiting assistant professor of environmental studies at the College of Wooster.

BEATRIZ CORTEZ is a visual artist and professor of Central American studies at California State University, Northridge. She is author of *Estética del cinismo: Pasión y desencanto en la literatura centroamericana de posguerra.*

WAI CHEE DIMOCK, editor of PMLA, teaches at Yale University. Among her books are *Through Other Continents: American Literature across Deep Time* and a team-edited anthology, *American Literature in the World.* Her latest book is *Weak Planet: Literature and Assisted Survival.*

JANE E. DMOCHOWSKI is a senior lecturer of earth and environmental science at the University of Pennsylvania. She is author of multiple peer-reviewed publications and has taught earth and environmental science courses at four colleges and universities.

DAVID A. D. EVANS is professor of earth and planetary sciences at Yale University. A fellow of the Geological Society of America, he is author of more than one hundred peer-reviewed publications and co-editor of the books *Palaeoproterozoic Supercontinents and Global Evolution* and *Supercontinent Cycles through Earth History.*

KATE FARQUHAR is a Philadelphia-based landscape designer at OLIN and has worked at the intersection of ecology, infrastructure, and art for fifteen years.

MARCIA FERGUSON is a senior lecturer in the Theatre Arts Program at the University of Pennsylvania. She is author of *Blanka and Jiri Zizka at the Wilma Theater, 1979–2000: From the Underground to the Avenue* and *A Short Guide to Writing about Theatre.*

CAROLYN FORNOFF is assistant professor of Latin American cultures at the University of Illinois at Urbana-Champaign. She is coeditor of *Pushing Past the Human in Latin American Cinema.*

ÖMÜR HARMANŞAH is associate professor of art history at the University of Illinois at Chicago. He is author of *Cities and the Shaping of Memory in the Ancient Near East* and *Place, Memory, and Healing: An Archaeology of Anatolian Rock Monuments.*

TROY HERION is a composer for film, theater, dance, and experimental arts.

PATRICIA EUNJI KIM is an art historian of the ancient Mediterranean and Middle East. She is assistant professor/faculty fellow at the Gallatin School of Individualized Study and a Provost's Postdoctoral Fellow at New York University. She also serves as a curator at Monument Lab, a public art and history studio.

MIMI LIEN is a designer of sets/environments for theater, dance, and opera.

MARY MATTINGLY is a social practice artist based in New York City. She is the founder of Swale, an edible landscape on a barge that circumvents public land laws, and creator of *WetLand,* a floating sculpture. She is artist in residence at the Brooklyn Public Library.

PAUL MITCHELL is a graduate student in anthropology at the University of Pennsylvania.

FRANK PAVIA is an isotope geochemist and oceanographer at the California Institute of Technology.

DAN ROTHENBERG, co-artistic director of Pig Iron Theatre Company, is a director and creator of experimental performance. Pig Iron, based in Philadelphia, has created more than thirty works of original performance, with tours across the United States and to fourteen countries.

JENNIFER E. TELESCA is assistant professor of environmental justice in the Department of Social Science and Cultural Studies at the Pratt Institute. She is author of *Red Gold: The Managed Extinction of the Giant Bluefin Tuna* (Minnesota, 2020).

CHARLES M. TUNG is professor of English at Seattle University, where he teaches courses on twentieth- and twenty-first-century literature, temporal scale, and representations of racial anachronism. He is the author of *Modernism and Time Machines.*

BETHANY WIGGIN is associate professor of German and founding director of the Penn Program in Environmental Humanities at the University of Pennsylvania. Her work explores changes in the land and waterscapes across the North Atlantic. She is at work on *Utopia Found and Lost in Penn's Woods,* an exploration of the end of worlds and a meditation on historical method amid ecological crisis.